T0276152

CAMBRIDGE LIBRARY COLLECTION

Books of enduring scholarly value

Technology

The focus of this series is engineering, broadly construed. It covers technological innovation from a range of periods and cultures, but centres on the technological achievements of the industrial era in the West, particularly in the nineteenth century, as understood by their contemporaries. Infrastructure is one major focus, covering the building of railways and canals, bridges and tunnels, land drainage, the laying of submarine cables, and the construction of docks and lighthouses. Other key topics include developments in industrial and manufacturing fields such as mining technology, the production of iron and steel, the use of steam power, and chemical processes such as photography and textile dyes.

Practical Essay on the Strength of Cast Iron and Other Metals

Although cast iron was used in pagoda construction in ancient China, it was in Britain in the eighteenth century that new methods allowed for its production in quantities that enabled widespread use. An engineer who had educated himself tirelessly in technical subjects from carpentry to architecture, Thomas Tredgold (1788–1829) first published this work in 1822. It served as a standard textbook for British engineers in the early nineteenth century, and several translations extended its influence on the continent. Reissued here in the fourth edition of 1842, edited and annotated by the structural engineer Eaton Hodgkinson (1789–1861), who presents his own research in the second volume, this work addresses both practical and mathematical questions in assessing metallic strength. In Volume 2, benefiting from twenty years of progress since Tredgold's original publication, Hodgkinson provides details of his own advanced experiments.

Cambridge University Press has long been a pioneer in the reissuing of out-of-print titles from its own backlist, producing digital reprints of books that are still sought after by scholars and students but could not be reprinted economically using traditional technology. The Cambridge Library Collection extends this activity to a wider range of books which are still of importance to researchers and professionals, either for the source material they contain, or as landmarks in the history of their academic discipline.

Drawing from the world-renowned collections in the Cambridge University Library and other partner libraries, and guided by the advice of experts in each subject area, Cambridge University Press is using state-of-the-art scanning machines in its own Printing House to capture the content of each book selected for inclusion. The files are processed to give a consistently clear, crisp image, and the books finished to the high quality standard for which the Press is recognised around the world. The latest print-on-demand technology ensures that the books will remain available indefinitely, and that orders for single or multiple copies can quickly be supplied.

The Cambridge Library Collection brings back to life books of enduring scholarly value (including out-of-copyright works originally issued by other publishers) across a wide range of disciplines in the humanities and social sciences and in science and technology.

Practical Essay on the Strength of Cast Iron and Other Metals

Containing Practical Rules, Tables, and Examples, Founded on a Series of Experiments, with an Extensive Table of the Properties of Materials

VOLUME 2:
EXPERIMENTAL RESEARCHES ON THE STRENGTH AND OTHER PROPERTIES OF CAST IRON

EATON HODGKINSON

CAMBRIDGE
UNIVERSITY PRESS

University Printing House, Cambridge, CB2 8BS, United Kingdom

Cambridge University Press is part of the University of Cambridge.
It furthers the University's mission by disseminating knowledge in the pursuit of education, learning and research at the highest international levels of excellence.

www.cambridge.org
Information on this title: www.cambridge.org/9781108070355

This edition first published 1846
This digitally printed version 2014

ISBN 978-1-108-07035-5 Paperback

EXPERIMENTAL RESEARCHES

ON THE

STRENGTH AND OTHER PROPERTIES

OF

CAST IRON:

WITH THE DEVELOPEMENT OF NEW PRINCIPLES;

CALCULATIONS DEDUCED FROM THEM;

AND

INQUIRIES APPLICABLE TO RIGID AND TENACIOUS BODIES GENERALLY.

BY EATON HODGKINSON, F.R.S.

WITH PLATES AND DIAGRAMS.

London:
JOHN WEALE, 59, HIGH HOLBORN.

M.DCCC.XLVI.

CONTENTS OF PART II.

PART II.

EXPERIMENTAL RESEARCHES

ON THE

STRENGTH AND OTHER PROPERTIES OF CAST IRON;

WITH

THE DEVELOPEMENT OF NEW PRINCIPLES; CALCULATIONS DEDUCED FROM THEM; AND INQUIRIES APPLICABLE TO RIGID AND TENACIOUS BODIES GENERALLY.

BY EATON HODGKINSON, F.R.S.

X

EXPERIMENTAL RESEARCHES, &c.

1. In the preceding Work the very ingenious Author has confined his reasonings chiefly to the effects produced upon bodies by forces, which were small comparatively to those necessary to produce fracture. In this Second, or Additional Part, I shall generally give the ultimate strength of the bodies experimented upon, and endeavour to show the laws, or illustrate the phenomena, attendant upon fracture. The conclusions in this Part will be drawn from experiments mostly made since Mr. Tredgold wrote his Work upon Cast Iron, at which time there was confessedly a want of experimental information upon the subject in this country ; and we were but slightly acquainted with what had been done upon the Continent. Having for many years devoted a portion of my time to experimental research on the strength of materials, in which the expense was borne by my liberal friend William Fairbairn, Esq., whose extensive mechanical establishment, at Manchester, enabled him to offer me every facility for the purpose, I have obtained a large mass of facts

on most of the subjects connected with the strength of materials. Mr. Fairbairn has likewise published the results of a great number of well-conducted experiments upon the transverse strength of bars of cast iron. An abstract, therefore, of the leading experiments made at Mr. Fairbairn's Works, and of those given by Navier, Rennie, Bramah, and others, with theoretical considerations, is all that can be attempted in this Additional Part; pointing out, as I proceed, whatever has a bearing upon the conclusions of Tredgold in the body of the Work.

TENSILE STRENGTH OF CAST IRON.

2. To determine the direct tensile strength of cast iron, I had models made of the form in Plate I. fig. 1.

The castings from these models were very strong at the ends, in order that they might be perfectly rigid there, and had their transverse section, for about a foot in the middle, of the form in fig. 2. This part, which was weaker than the ends, was intended to be torn asunder by a force acting perpendicularly through its centre. The ends of the castings had eyes made through them, with a part more prominent than the rest in the middle of the casting, where the eye passed through; fig. 3 represents a section of the eye. The intention of this was that bolts passing through the eyes, and having shackles attached to them, by which to tear the casting asunder, would rest upon this prominent part in the middle, and therefore upon a point passing in a direct line through the axis of the casting. Several of the castings were torn asunder by the machine for testing iron cables, belonging to the Corporation of Liverpool; the results from these are marked with an asterisk. Others were made in the same manner, but of smaller transverse area; these were broken by means of Mr. Fairbairn's lever (Plate II. fig. 40), which was adapted so as to be well suited for the purpose.

The form of casting here used was chosen to obviate the objections made by Mr. Tredgold (art. 79 and 80) and others against the conclusions of former experimenters. The results are as follow:

3. *Results of Experiments on the Tensile Strength of Cast Iron.*

Description of iron.	Area of section in inches.	Breaking weight in lbs.	Strength per sq. in. of section.	Mean in lbs. per square inch.
				tons. cwt.
Carron iron, No. 2, hot blast	4·031	56,000	13,892*	
Do. do. do.	1·7236	22,395	12,993	13,505 = 6 0½
Do. do. do.	1·7037	23,219	13,629	
Carron iron, No. 2, cold blast	1·7091	28,667	16,772	16,683 = 7 9
Do. do. do.	1·6331	27,099	16,594	
Carron iron, No. 3, hot blast	1·7023	28,667	16,840	17,755 = 7 18½
Do. do. do.	1·6613	31,019	18,671	
Carron iron, No. 3, cold blast	1·6232	22,699	13,984	14,200 = 6 7
Do. do. do.	1·6677	24,043	14,417	
Devon (Scotland) iron, No. 3, hot blast	4·269	93,520	21,907*	21,907 = 9 15½
Buffery iron, No. 1, hot blast	3·835	51,520	13,434*	13,434 = 6 0
Buffery iron, No. 1, cold blast	4·104	71,680	17,466*	17,466 = 7 16
Coed-Talon (North Wales) iron, No. 2, hot blast . .	1·586	25,818	16,279	16,676 = 7 9
Do. do. do.	1·645	28,086	17,074	
Coed-Talon (North Wales) iron, No. 2, cold blast . .	1·535	30,102	19,610	18,855 = 8 8
Do. do. do.	1·568	28,380	18,100	
Low Moor iron (Yorkshire), No. 3, from 5 experiments	1·540	22,385	14,535	14,535 = 6 10
Mixture of iron,—4 experiments further on (art. 7.)				16,542 = 7 7¾
Mean from the whole				16,505 = 7 7⅓

4. The preceding Table, excepting the two last lines, is extracted from my Report on the strength and other properties of cast iron obtained by hot and cold blast, in vol. vi. of the British Association.

5. In the second volume published by the Asso-

ciation, there are given the results of a few experiments, which I made to determine the tensile strength of cast iron of the following mixture: Blaina No. 2 (Welsh), Blaina No. 3, and W. S. S., No. 3 (Shropshire), each in equal portions.[1]

6. In these experiments the intention was to determine, first, the direct tensile strength of a rectangular mass, when drawn through its axis, and next the strength of such a mass, when the force was in the direction of its side. The castings for the experiments on the central strain were of the form previously described; and in the others the force was exactly along the side. The experiments were made by means of the Liverpool testing machine.

7. *Force up the middle.*

Experiments.	Area of section in parts of an inch.	Breaking weight in tons.	Strength per square inch.
1	3·012	22·5	7·47 ⎫
2	2·97	21·0	7·07 ⎬ mean 7·65 tons. = 17136 lbs.
3	3·031	25·5	8·41 ⎭
4	2·95	19·5	6·59 different quality of iron.

[1] This mixture of iron is the same as I had employed in some experiments on the strength and best forms of cast iron beams, (Memoirs of the Literary and Philosophical Society of Manchester, vol. v., second Series,) of which an account will be given further on.

8. *Force along the side.*

Experiments.	Area of section in parts of an inch.	Breaking weight in tons.	Strength per square inch.
5	4·83	11·5	2·38 } mean 2·62 tons.
6	4·815	13·75	2·855 }

9. Whence we see that the strength of a rectangular piece of cast iron, drawn along the side, is about one-third, or a little more, of its strength to resist a central strain, since $3 \times 2{\cdot}62 = 7{\cdot}86$ is somewhat greater than 7·65. Mr. Tredgold computed that, if the elasticity remained perfect, the strength in these two cases would be as 4 to 1 (art. 281).

10. The following Table, calculated from one given by Navier (Application de la Mécanique), contains the results of several experiments made in 1815 by Minard and Desormes upon the direct tensile strength of cylindrical pieces of cast iron, of which the specific gravity was 7·074. The results from the experiments which they have given on defective specimens are here rejected.

No. of experiments	Temperature.		Area of transverse section.	Weight producing rupture.		
				Total.	Per square millimètre.	Per English inch square.
	Degrees of centrigade thermometer.	Degrees of Fahrenheit.	Square millimètres.	Kilogrammes.	Kilogrammes.	Tons. English.
1	− 6	21·2	330	3392	10·2788	6·5277
2	− 5	23	346	3542	10·2370	6·5011
3	− 5	23	363	3092	8·5179	5·4094
4	−15	5	363	3720	10·2479	6·5080
5	+60	140	353	4020	11·3881	7·2321
6	+ 3	37·4	147	1920	13·0612	8·2946
7	+ 5	41	165	1920	11·6364	7·3898
8	+ 5	41	165	2140	12·9691	8·2362
9	+ 5	41	165	2360	14·3030	9·0833
13	+ 5	41	346	3670	10·6069	6·7360

11. Since a square millimètre is ·001550059 of an English square inch, and a kilogramme = 2·205 of a pound avoirdupois, multiplying the kilogramme per square millimètre, in the last column but one of the Table above, by 2·205, and dividing the product by ·001550059, and by 2240, the number of pounds in a ton, gives the number of tons per square inch which the iron required to tear it asunder. Or if we multiply the numbers in the last column but one by ·63506, we obtain the same result; and thus the last column was formed.

12. We find from these experiments that the strength of the weakest specimen was 5·09 tons per square inch, that of the strongest 9·08 tons, and the mean strength from all the specimens was 7·19 tons.

13. In my own experiments given above, in which every care was taken both to form the cast-

ings in such a manner as to obviate theoretical objections, and to obtain accurate results, the strengths varied from 6 tons to 9¾ tons per square inch, the mean from twenty-five experiments being 16,505 lbs. or 7·37 tons nearly. These experiments were upon iron obtained from various parts of England, Scotland, and Wales; and in no case, except one, was it found to bear more than 8½ tons per square inch. With these facts before the reader, he will, I conceive, be unable to see how the mean tensile strength of cast iron can properly be assumed at more than 7 or 7½ tons per square inch; but some of our best writers have, by calculating the tensile strength from experiments on the transverse, arrived at the conclusion that the strength of cast iron is 10, or even 20 tons, or more. Mr. Barlow conceives it to be upwards of 10 tons (Treatise on Strength of Timber, Cast Iron, &c., p. 222), and Mr. Tredgold makes it at least 20 (art. 72 to 76). The reasoning of Mr. Tredgold, by which he arrives at this erroneous conclusion, with others resulting from it, will be examined at length under the head "Transverse Strength." Navier, too, (Application de la Mécanique, article 4,) calculates the tensile strength of cast iron from principles somewhat similar to those assumed by Tredgold, and finds it much too high.

STRENGTH OF CAST IRON AND OTHER MATERIALS TO RESIST COMPRESSION.

14. On this subject there was acknowledged to exist a greater want of experimental research than on any other connected with the strength of materials. Feeling this to be the case, I have done all that I could, without too lavish a use of the means intrusted to me, to supply the deficiency.

15. The matter will be classed under two heads. 1st. The resistance of bodies which are so short, compared with their lateral dimensions, as to be crushed with little or no flexure. 2nd. The resistance of pillars long enough to break by flexure.

RESISTANCE OF SHORT MASSES TO A CRUSHING FORCE.

16. On this subject I shall give, as before, an abstract from my Report on the strength and other properties of hot and cold blast iron, in the sixth volume of the British Association, referring for more information to the Work itself. Great diversity exists in the conclusions of former experimenters upon the matter. Rondelet found (Traité de l'Art de bâtir) that cubes of malleable iron, and prisms of various kinds of stone, were crushed with forces which were directly as the area, whilst from Mr. Rennie's experiments, both upon cast iron and wood, it would appear that the resistance increases, particularly in the latter, in a much higher ratio than the area (Mr. Barlow's Treatise on the Strength

of Materials, art. 112). With respect to M. Rondelet's conclusions, that cubes of malleable iron resisted crushing with forces proportional to their areas, and that to such a degree, that when in his experiments the area was increased four times, this ratio did not differ from the result so much as a fiftieth part, I am strongly persuaded that *wrought* iron does not admit of such precision of judging when crushing commences, as to enable any conclusion to be easily drawn with respect to its proportionate resistance to crushing. A prism of that metal becomes slightly flattened and enlarged in diameter with about 9 or 10 tons per square inch, and this effect increases as the weight is increased; but there is no abrupt change in the metal by disunion of the parts, as in cast iron, wood, &c.

17. With respect to the experiments of Mr. Rennie, the lever used in performing them would not, during its descent, act uniformly upon all parts of the specimen; and therefore the results would be liable to objection. I endeavoured therefore, by repeating, with considerable variations, in the Report above named, the ingenious experiments of Mr. Rennie, to arrive at some definite conclusions upon the subject.

18. In order to effect this, it was thought best to crush the object between two flat surfaces, taking care that these were kept perfectly parallel, and that the ends of the prism to be crushed were turned parallel, and at right angles to their axes;

so that when the specimen was placed between the crushing surfaces its ends might be completely bedded upon them. For this purpose a hole, $1\frac{1}{4}$ inch diameter, was drilled through a block of cast iron, about 5 or 6 inches square, and two steel bolts were made which just fitted this hole, but passed easily through it; the shortest of these bolts was about $1\frac{1}{4}$ inch long, and the other about 5 inches; the ends of these bolts were hardened, having previously been turned quite flat and perpendicular to their axes, except one end of the larger bolt, which was rounded. The specimen was crushed between the flat ends of these bolts, which were kept parallel by the block of iron in which they were inserted. See fig. 4, where A and B represent the bolts, with the prism C between them, and D E the block of iron. During the experiment the block and bolt B rested upon a flat surface of iron, and the rounded end of the bolt A was pressed upon by the lever. There was another hole drilled through the block at right angles to that previously described; this was done in order that the specimen might be examined during the experiment, and previous to it, to see that it was properly bedded.

19. The specimens were crushed by means of the lever represented in Plate II. fig. 40, the bolt A (Plate I. fig. 4) being placed under it in the manner the pillar is there described to be. In the experiments the lever was kept as nearly horizontal as possible.

The results of the first experiments I made are given in the following Table:

20. *Tabulated results of experiments made to ascertain the weights necessary to crush given cylinders, &c. of cast iron, of the quality No. 2, from the Carron Iron Works. The specimens in the first three columns of results were from cylinders cast for the purpose, and turned to the right size; the ½ inch from those of about ¾ inch, &c.*

Height of specimens.	Cylinder ¼ inch diameter, area of base ·0491. Crushing weight.	Cylinder ⅜ inch diameter, area of base ·1104. Crushing weight.	Cylinder ½ inch diameter, area of base ·1963. Crushing weight.	Cylinder ·64 inch diameter, area of base ·3217. Crushing weight.	Right prisms, base an equilateral triangle, circumscribing an ½ inch cylinder, its sides being ·866 in., area ·3247. Crushing weight.	Right prisms, bases squares, ½ inch the side, cut out of an inch square bar, area ·250. Crushing weight.	Right prisms, base a rectangle 1·00 × ·251 inch, cut out of a bar 1¼ inch square, area ·251 inch. Crushing weight.
	lbs.	lbs.	lbs.	lbs.			
⅛ inch.	{ 8737 8145 }	18,882	30,461				
¼ do.	6818	16,474	26,983				
⅜ do.	6563	13,736	26,412				
½ do.	6301	13,638	24,210		35,548	25,721	27,187
⅝ do.	6309	14,156		38,671			
¾ do.	5980	15,059	23,465	35,888	33,448	24,191	
⅞ do.				35,888			
1 do.	5798	14,877	22,867		31,348	23,950	25,151
1¼ do.		14,190	24,177				
1½ do.		14,143	23,453				
2 do.		13,800	21,828				

21. By comparing the results in each vertical column, we see that the shorter specimens generally bear more than the larger ones of the same diameter, or dimensions of base. In the shortest specimens fracture takes place by the middle becoming flattened and increased in breadth, so as to burst the surrounding parts, and cause them to be crumbled and broken in pieces. This is usually the case when the lateral dimensions of the prism are large compared with the height. When they are equal to, or less than the height, fracture is caused by the body becoming divided diagonally in one or more directions. In this case the prism, in cast iron at least, either does not bend before fracture, or bends very slightly; and therefore the fracture takes place by the two ends of the specimen forming cones or pyramids, which split the sides and throw them out; or, as is more generally the case in cylindrical specimens, by a wedge sliding off, starting at one of the ends, and having the whole end for its base; this wedge being at an angle which is constant in the same material, though different in different materials. In cast iron the angle is such that the height of the wedge is somewhat less than $\frac{3}{2}$ of the diameter. The forms of fracture in these cases may be seen from Plate I., in which fig. 5 represents a cylinder before fracture, and fig. 6 the same cylinder afterwards; a part in the form of a wedge having separated from one side of it, and the remainder being shortened and bulged out in the

middle, which is very obvious in experiments on soft cast iron. Figs. 7 and 8 represent another cylinder before and after fracture: in fig. 8 the sides are separated by the action of two cones, having the ends of the cylinder for the bases, and the vertices with sharp edges or points formed near to the centre of the cylinder, but inclined a little from the axis, so that they may slip past each other, and divide the mass without injury to the cones. Figs. 9 and 10 show the same thing as the two preceding figures; and fig. 11 is a representation of one of the cones, the vertex being sharp, as above mentioned. Fig. 12 is another cylinder, of soft cast iron, showing the directions of the fractures, but not separated. Fig. 13 was a cylinder too short to be crushed in the ordinary way; but the centre shows the rudiments of a cone, throwing out the surrounding parts. Fig. 14 represents a rectangular prism; and figs. 15 and 16 the modes of its fracture. Figs. 17 and 18 represent a whole and fractured prism; and the same is the case with respect to figs. 19 and 20. In fig. 20 the sharp-pointed pyramid, with the lower side of the specimen for its base, is very clearly shown; it has cut up the prism, separating the sides, and left a number of sharp-edged parts, all of which have slid off in the angle of least resistance. Figs. 21 and 22 exhibit the appearance of a triangular prism, before and after fracture; two pyramids, formed as usual, with their bases at the ends, and the vertices toward the centre, have thrown out the angular parts. The parts, so separated, have in prisms of every form a general resemblance, and the form of the pyramidal wedge has considerable interest, as it is that of least resistance in cast iron, and furnishes hints as to the best forms of cutters for that metal. Further investigations upon the subject of this article, of a theoretical nature, will be given on a future occasion.

22. The mode of fracture is the same, and the strength of the specimen very little diminished, by any increase of its height, whilst its lateral dimensions are the same, provided the height be greater than the diameter, when the body is a cylinder, but not greater, in cast iron, than four or five times the diameter, or least lateral dimension in specimens not cylindrical. If the length be greater than this, the body bends with the pressure, and though it may break by sliding off as before, the strength is much decreased. In cases where the length is much greater than as above, the body breaks across, as if bent by a transverse pressure.

23. The preceding Table was formed by taking the means from results on the resistance to crushing of specimens of equal size in Carron Iron, No. 2, one Table being on hot blast iron, and the other cold; the mean from the results being given here in one Table, in preference to the two Tables at large, as in the volume of the Association; since most of the results thus obtained are means from

several experiments; and there was very little difference in the strength of the hot and cold blast iron of this description to resist crushing.

24. Comparing the results in the different vertical columns of the Table, it appears that the strength of the specimens was nearly as the area of their transverse section. Thus the cylinders ½ inch diameter bore nearly four times as much as those of ¼ inch. The falling off from this proportion in the strength of some of the larger specimens, I attribute to those having been cut out of larger (and consequently softer) masses of the iron than the small specimens.

25. To obtain further evidence on the matter I crushed in the same manner twelve right cylinders of Teak-wood, ½ inch, 1 inch, and 2 inches diameter, four of each, the latter eight out of the same piece of wood; the height in each case was double the diameter. The pressure was in the direction of the fibres. The strengths were as below.

Cylinders ½ inch diameter.		Cylinders 1 inch diameter.		Cylinders 2 inches diameter.	
2335	mean 2439 lbs.	10507	mean 10171 lbs.	38909	mean 40304 lbs.
2543		9499		39721	
2543		10507		41294	
2335		10171		41294	

26. The mean crushing weights above are nearly as 25, 100, and 400, which is the ratio of the areas of the sections of the cylinders. The strengths are therefore as the areas, though these vary as 4 and 16 to 1.

27. In this and every other kind of timber, like as in iron and crystalline bodies generally, crushing takes place by wedges sliding off in angles with their base, which may be considered constant in the same material: hence the strength to resist crushing will be as the area of fracture, and consequently as the direct transverse area; since the area of fracture would, in the same material, always be equal to the direct transverse area, multiplied by a constant quantity.

28. In estimating the resistance, per square inch, of the iron above to a crushing force, I shall mostly confine myself to such specimens as vary in height from about the length of the wedge, which would slide off to twice its length,—say, such as have their height from $1\frac{1}{2}$ to 3 times the diameter,—thus avoiding the results from prisms which, through their shortness, could not break by a wedge sliding off at its proper angle, and therefore must offer an increased resistance; and those whose extra length would enable them to bend before sliding off, and thus have their strength reduced. I shall here take the results from my experiments upon the hot and cold blast iron separately, that the difference may be seen; the iron being of the same materials in the two cases, and affected only by the heat of the blast used in its preparation.

The results are in the following Tables.

29. *Resistance of Hot Blast Iron (Carron, No. 2) to a crushing force.*

Dimensions of base of specimen.	Number of experiments derived from.	Mean crushing weight.	Mean crushing weight per square inch.	General mean per square inch.
		lbs.	lbs.	
Right cylinder, diameter $\frac{1}{4}$ inch	3	6,426	130,909	From the cylinders 121,685 lbs. =54 tons $6\frac{1}{2}$ cwts.
,, $\frac{3}{8}$,,	4	14,542	131,665	
,, $\frac{1}{2}$,,	5	22,110	112,605	
,, $\frac{16}{25}$=·64	1	35,888	111,560	
Right prism, area ·50 × ·50 in.	3	25,104	100,416	From the prisms 100,739 lbs. =44 tons $19\frac{1}{2}$ cwts.
Do. area 1·00 × ·26	2	26,276	101,062	

Mean per sq. in. from the whole of the experiments, 114,703 lbs. =51 tons 4 cwts.

30. *Resistance of Cold Blast Iron (Carron, No. 2) to a crushing force.*

Dimensions of base of specimen.	Number of experiments derived from.	Mean crushing weight.	Mean crushing weight per square inch.	General mean per square inch.
		lbs.	lbs.	
Right cylinder, diameter ¼ inch	2	6,088	124,023	From the cylinders 118,211 lbs. = 52 tons 15½ cwts.
„ ⅜ „	4	14,190	128,478	
„ ½ „	7	24,290	123,708	
„ ·45 „ and ·75 high	2	15,369	96,634	
Equilateral triangle, side ·866 inch	2	32,398	99,769	From the prisms 101,964 lbs. = 45 tons 10½ cwts.
Square, area ·50 × ·50 inch	2	24,538	98,152	
Rectangle, area 1·00 × ·243	3	26,237	107,971	
Mean per sq. in. from the whole of the experiments, 111,248 lbs. = 49 tons 13¼ cwts.				

31. Comparing the results in these two Tables, it will be seen, as has been mentioned before, that the Carron Iron, No. 2, offers but little difference of resistance to a crushing force, whether the iron be prepared with a hot or a cold blast. The falling off in the resistance per square inch in the latter class of experiments, in each Table, compared with the former, has been attempted to be accounted for by the iron out of which the larger specimens were

cut being softer and weaker than the thin cylinders out of which the smaller specimens were obtained.

32. The resistances of other species of cast iron to a crushing force, obtained in the same manner as above, are as in the following Table.

Description of iron.	Form of specimen.	Number of experiments derived from.	Mean strength per square inch.
			lbs. tons. cwts.
Devon (Scotch) iron, No. 3, hot blast	Cylinder.	4	145,435 = 64 18½
Buffery (near Birmingham) iron, No. 1, hot blast . .	Do.	4	86,397 = 38 11¼
Do., cold blast	Do.	4	93,385 = 41 13½
Coed-Talon (Welsh) iron, No. 2, hot blast . . .	Do.	4	82,734 = 36 18½
Do., cold blast	Do.	4	81,770 = 36 10
Carron (Scotch) iron, No. 2, hot blast	Cylinders and prisms.	18	114,703 = 51 4
Do., cold blast	Do.	22	111,248 = 49 13¼
Carron iron, No. 3, hot blast	Prisms.	3	133,440 = 59 11¼
Do., cold blast	Do..	4	115,442 = 51 10¾
			mean. tons. cwts.
Low Moor (Yorkshire) iron, No. 3, cold blast . . .	Cylinder. Rectangle.	3 2	115,911 103,692 } 109,801 = 49 0⅓
Mixture of iron, same as in my experiments on the strength of beams (see note to art. 5)	Cylinders ·508 and ·6 inch. diameter, 3 of each. Rectangles cut out of a beam.	6 3	100,049 121,767 } 110,908 = 49 10¼

33. The ratios of the forces necessary to crush and tear asunder equal masses of cast iron may now be obtained; the experiments of which I have given

the results, in this and the preceding article, will supply those ratios which are in the following Table.

Description of metal.	Crushing force per square inch in lbs.	Tensile force per square inch in lbs.	Ratio.	
Devon iron, No. 3, hot blast . .	145,435	21,907	6·638 : 1	Mean 6·595 : 1
Buffery iron, No. 1, hot blast . .	86,397	13,434	6·431 : 1	
Do. No. 1, cold blast .	93,385	17,466	5·346 : 1	
Coed-Talon iron, No. 2, hot blast	82,734	16,676	4·961 : 1	
Do. No. 2, cold blast	81,770	18,855	4·337 : 1	
Carron iron, No. 2, hot blast . .	114,703	13,505	8·493 : 1	
Do. No. 2, cold blast .	111,248	16,683	6·668 : 1	
Do. No. 3, hot blast . .	133,440	17,755	7·515 : 1	
Do. No. 3, cold blast .	115,442	14,200	8·129 : 1	
Low Moor iron, No. 3, cold blast	109,801	14,535	7·554 : 1	
Mixture of iron used in my experiments on beams (art. 5-7)	110,908	17,136	6·472 : 1	

34. From this Table it appears that cast iron requires from 4·337 to 8·493 times as much force to crush it as to tear it asunder, the mean being 6·595. I conceive, however, this mean to be too low; it would have been between 7 and 8 if the prisms which were crushed had been cut out of the same masses as those which were torn asunder; but they were in many instances out of larger castings, which, being softer, offered less resistance.

STRENGTH OF LONG PILLARS.

35. I shall here consider such pillars as are too long to break by sliding off, as in the last article. This class will include all such as are usually made of iron or timber, the lengths of the shortest of which are generally many times their lateral dimensions; and it has been remarked before, that in some cases right cylinders of cast iron, whose length did not exceed 5 times the diameter, became bent under a direct force of compression, so as to break nearly straight across in the manner of longer columns.

The acknowledged want of practical information upon this subject,[2] and its great importance, made me anxious to undertake an extensive series of experiments upon it, such as would confirm or show the error of existing theories, and give such information as would be of real service to the engineer and architect, whilst they tended to unfold the laws that regulate the strength of pillars. This wish was, as on other occasions, cheerfully responded to by my friend William Fairbairn, Esq., at whose expense the extensive series of experiments, of which the following is an abstract,[3] was made.

---

[2] Dr. Robison (Mechanical Philosophy, vol. i.) and Mr. Barlow (Report to the British Association) have strongly expressed their opinion of our want of information upon it; and the British Association have mentioned such experiments as among the desiderata of Practical Science.

[3] See Experimental Researches on the Strength of Pillars of

36. In the original Paper the experiments are contained in thirteen Tables,[4] as below.

	CAST IRON.	No. of experiments.
Table I.	Solid uniform cylindrical pillars, with rounded ends (fig. 23)	55
II.	Do., with flat ends (fig. 26)	65
III.	Solid uniform square pillars, with rounded ends (fig. 28)	7
IV.	Solid uniform cylindrical pillars, with discs (fig. 27)	12
V.	Do., with ends rounded (fig. 23), rounded and flat (fig. 24), and both ends flat (fig. 26) .	23
VI.	Solid cylindrical pillars, enlarged middle, rounded ends (fig. 31)	7
VII.	Do., do., discs on ends (fig. 32)	4
VIII.	Hollow uniform cylinders, rounded ends (fig. 36)	19
IX.	Do., with flat ends (fig. 37)	11
X.	Short hollow uniform cylinders, with flat ends (fig. 37)	13
XI.	Pillars, hollow and solid, of various forms, and different modes of fixing (figs. 33, 34, 35, 38, 39)	10
	WROUGHT IRON AND STEEL.	
XII.	Solid uniform cylindrical pillars of these metals (figs. 23, 24, 25, 26, 27)	30
	WOOD.	
XIII.	Pillars of oak and red deal, square and other rectangular forms (figs. 28, 29, 30, &c.)	21
	Number of experiments	277

Cast Iron and other Materials.—Phil. Trans. of the Royal Society, Part II. 1840.

[4] Abstracts of Tables I., II., VIII., IX., X., are given in the present Work.

37. The drawing, Plate II. fig. 40, will show how the experiments were made. The pillars were placed vertically, resting upon a flat smooth plate of hardened steel, laid upon a cast iron shelf, E, made very strong, and lying horizontally. The pressure was communicated to the upper end of the pillar by means of the lever, A B, acting upon a bolt, C, of hardened steel, 2½ inches diameter, and about a foot long, kept vertical by being made to pass through a hole bored in a deep mass, D, of cast iron, the hole being so turned as just to let the bolt slide easily through without lateral play. The top of the bolt was hemispherical, that the pressure from the lever might act through the axis of it; and the bottom was turned flat to rest upon the pillar, I K. The bottom of this bolt, and the shelf on which the pillar stood, were necessarily kept parallel to each other; for the mass through which the bolt passed, and that on which the shelf rested, were parts of the same large case, D F G, of iron, cast in one piece, and so formed as to admit shelves at various heights for breaking pillars of different lengths. The case had three of its four sides closed; circular apertures were, however, made through them, that the experimenter might observe the pillar without danger.

38. With a view to develope the laws connecting the strength of cast iron pillars with their dimensions, they were broken of various lengths, from 5 feet to 1 inch, and their diameters varied from

$\frac{1}{2}$ an inch to 2 inches in solid pillars; and in hollow ones the length was increased to 7 feet 6 inches, and the diameter to $3\frac{1}{2}$ inches. My first object was to supply the deficiencies of Euler's theory of the strength of pillars (Académie de Berlin, 1757,) if it should appear capable of being rendered practically useful; and if not, to endeavour to adapt the experiments so as to lead to useful conclusions. As the results of the experiments were intended to be compared together, it was desirable that all the pillars of cast iron should be from one species of metal; and the description chosen was a Yorkshire iron, the Low Moor, No. 3, which is a good iron, not very hard, and differs not widely from that called No. 2. The pillars were mostly made cylindrical, as that seemed a more convenient form in experiments of this kind than the square; for square pillars do not bend or break in a direction parallel to their sides, but to their diagonals, nearly. The experiments in the first Table were made on solid uniform pillars rounded at the ends to render them capable of turning easily there, and that the force might be through the axis; and in the second Table the pillars were uniform and cylindrical, as before, but had their ends flat and at right angles to the axis; so that the whole end of the pillar might be pressed upon, instead of the axis only, as in the last case; thus rendering the pillar incapable of moving at the ends. In the fifth Table (art. 36) uniform pillars, with one end rounded and one flat, were used; and to prove the constant connexion (in any particular material) between the results of these three kinds of pillars, all of equal dimensions, enters largely into the other Tables, which are occupied likewise in inquiries into the strength of hollow cylindrical pillars, and others of different forms.

39. RESULTS.

1st. In all long pillars of the same dimensions, the resistance to fracture by flexure is about three times greater when the ends of the pillar are flat and firmly bedded, than when they are rounded and capable of turning.[5]

2nd. The strength of a pillar, with one end round and the other flat, is the arithmetical mean between that of a pillar of the same dimensions with both ends rounded, and with both ends flat. Thus, of three cylindrical pillars, all of the same length and diameter, the first having its ends rounded, the second with one end rounded and one flat, and the third with both ends flat, the strengths are as 1, 2, 3, nearly.[6]

3rd. A long uniform pillar, with its ends firmly fixed, whether by discs or otherwise, has the same

[5] This will be seen by comparing the results from the longer pillars of equal size in Tables I. and II., page 329, of which abstracts are given further on.

[6] This was proved by Table V., page 329, and the next conclusion was obtained by a like comparison of results.

power to resist breaking as a pillar of the same diameter, and half the length, with the ends rounded or turned so that the force would pass through the axis.

The preceding properties were found to exist in long pillars of steel, wrought iron, and wood.

4th. The experiments in Tables VI. and VII. (art. 36) show that some additional strength is given to a pillar by enlarging its diameter in the middle part; this increase does not, however, appear to be more than $\frac{1}{7}$th or $\frac{1}{8}$th of the breaking weight.

5th. The index of the power of the diameter, to which the strength of long pillars of cast iron, with rounded ends, is proportional, is 3·76, nearly; and 3·55 in those with flat ends; as appeared from means between the results of a great number of experiments;[7] or the strength of both may be taken as following the 3·6 power of the diameter, nearly.

6th. In cast iron pillars of the same thickness the strength is inversely proportional to the 1·7 power of the length, nearly.[8]

40. Thus the strength of a solid pillar of that

---

[7] The pillars with rounded ends, compared together, varied in diameter (or side of square) from $\frac{1}{2}$ inch to 2 inches. There were seventeen of them, and the largest exponent of the diameter to which the strength was in any case proportional, is 3·928, and the smallest 3·425. The pillars with flat ends, and those with discs, compared as above, were eleven, and their exponents varied from 3·922 to 3·412.

[8] The exponent 1·7 for the strength, according to the inverse power of the length is from experiments upon nineteen pillars, varying in length from 60·5 to 3·78 inches; the highest and lowest exponents being 1·914 and 1·424.

material, with rounded ends, the diameter of the pillar being d, and the length l, is as

$$\frac{d^{3.6}}{l^{1.7}}.$$

41. The breaking weights of solid cylindrical cast iron pillars, as appeared from the experiments, are nearly as below.

In solid pillars with their ends rounded, and moveable as above, fig. 23,

$$\text{Strength in tons} = 14{\cdot}9 \times \frac{d^{3.6}}{l^{1.7}}.$$

In solid pillars with their ends flat, and incapable of motion, fig. 26,

$$\text{Strength in tons} = 44{\cdot}16 \times \frac{d^{3.6}}{l^{1.7}};$$

where l is in feet, and d in inches.

In hollow pillars nearly the same laws were found to obtain; thus, if D and d be the external and internal diameters of a cast iron pillar, whose length is l, the strength of a hollow pillar of which the ends were moveable, fig. 36, (as in the connecting rod of a steam engine,) would be expressed by the formula below.

$$\text{Strength in tons} = 13 \times \frac{D^{3.6} - d^{3.6}}{l^{1.7}}$$

In hollow pillars, whose ends are flat, and firmly fixed (fig. 37, &c.) by discs or otherwise, I found from the results of numerous experiments as before,

$$\text{Strength in tons} = 44{\cdot}3 \times \frac{D^{3.6} - d^{3.6}}{l^{1.7}}$$

42. The co-efficients given above in the formulæ for the strength of solid pillars were obtained as below:—The 14·9 is the mean result from 18

pillars in Table I., art. 36, varying in length from 121 times the diameter down to 15 times. The 44·16, for pillars with flat ends, is similarly obtained from 11 pillars in Table II., varying in length from 78 to 25 times the diameter. Flat-ended pillars, shorter than 25 or 30 times the diameter, require a modification of the above rule for their strength ; as I found them to be crushed as well as bent by the pressure, and therefore to have their strength decreased: the mode I used in this case will be seen in arts. 45 to 51, further on. The co-efficients for the strength of hollow pillars were obtained in the same manner as those for solid ones; the 13 is the mean obtained from 19 experiments in Table VIII. upon hollow pillars, with rounded caps upon the ends, to make them moveable there like solid ones with rounded ends: this number is too low, as some of the first hollow pillars were bad castings. The 44·3, for those with flat ends, was obtained from 11 pillars in Table IX.; taking pillars only whose length was more than 25 times the external diameter, as in solid ones.

The Tables from which these results were derived are given, slightly abridged, further on. The formulæ for hollow pillars were obtained by adapting the results of Euler's theory to those of experiment, and were found to answer well when so altered. According to that theory the strength varies as $\frac{D^4-d^4}{l^2}$, (Poisson, Mécanique, vol. i., 2nd edition, art. 315).

43. As the above expressions for the strength of pillars contain fractional powers of the diameters and lengths, these may be taken from the Tables below; the first Table comprising the involved values in inches of most of the diameters in common use; and the second those of the lengths in feet.

TABLE I.—*Powers of Diameters.*

$1·0^{3·6} = 1$	$4·25^{3·6} = 182·89$	$6·8^{3·6} = 993·19$
$1·25^{3·6} = 2·2329$	$4·3^{3·6} = 190·76$	$6·9^{3·6} = 1046·8$
$1·5^{3·6} = 4·3045$	$4·4^{3·6} = 207·22$	$7·0^{3·6} = 1102·4$
$1·75^{3·6} = 7·4978$	$4·5^{3·6} = 224·68$	$7·1^{3·6} = 1160·2$
$2·0^{3·6} = 12·125$	$4·6^{3·6} = 243·18$	$7·2^{3·6} = 1220·1$
$2·1^{3·6} = 14·454$	$4·7^{3·6} = 262·76$	$7·25^{3·6} = 1250·9$
$2·2^{3·6} = 17·089$	$4·75^{3·6} = 272·96$	$7·3^{3·6} = 1282·2$
$2·25^{3·6} = 18·529$	$4·8^{3·6} = 283·44$	$7·4^{3·6} = 1346·6$
$2·3^{3·6} = 20·055$	$4·9^{3·6} = 305·28$	$7·5^{3·6} = 1413·3$
$2·4^{3·6} = 23·3755$	$5·0^{3·6} = 328·32$	$7·6^{3·6} = 1482·3$
$2·5^{3·6} = 27·076$	$5·1^{3·6} = 352·58$	$7·7^{3·6} = 1553·7$
$2·6^{3·6} = 31·182$	$5·2^{3·6} = 378·10$	$7·75^{3·6} = 1590·3$
$2·7^{3·6} = 35·720$	$5·25^{3·6} = 391·36$	$7·8^{3·6} = 1627·6$
$2·75^{3·6} = 38·159$	$5·3^{3·6} = 404·94$	$7·9^{3·6} = 1704·0$
$2·8^{3·6} = 40·716$	$5·4^{3·6} = 433·13$	$8·0^{3·6} = 1782·9$
$2·9^{3·6} = 46·199$	$5·5^{3·6} = 462·71$	$8·25^{3·6} = 1991·7$
$3·0^{3·6} = 52·196$	$5·6^{3·6} = 493·72$	$8·5^{3·6} = 2217·7$
$3·1^{3·6} = 58·736$	$5·7^{3·6} = 526·20$	$8·75^{3·6} = 2461·7$
$3·2^{3·6} = 65·848$	$5·75^{3·6} = 543·01$	$9·0^{3·6} = 2724·4$
$3·25^{3·6} = 69·628$	$5·8^{3·6} = 560·20$	$9·25^{3·6} = 3006·85$
$3·3^{3·6} = 73·561$	$5·9^{3·6} = 595·75$	$9·5^{3·6} = 3309·8$
$3·4^{3·6} = 81·908$	$6·0^{3·6} = 632·91$	$9·75^{3·6} = 3634·3$
$3·5^{3·6} = 90·917$	$6·1^{3·6} = 671·72$	$10·0^{3·6} = 3981·07$
$3·6^{3·6} = 100·62$	$6·2^{3·6} = 712·22$	$10·25^{3·6} = 4351·2$
$3·7^{3·6} = 111·05$	$6·25^{3·6} = 733·11$	$10·5^{3·6} = 4745·5$
$3·75^{3·6} = 116·55$	$6·3^{3·6} = 754·44$	$10·75^{3·6} = 5165·0$
$3·8^{3·6} = 122·24$	$6·4^{3·6} = 798·45$	$11·0^{3·6} = 5610·7$
$3·9^{3·6} = 134·23$	$6·5^{3·6} = 844·28$	$11·25^{3·6} = 6083·4$
$4·0^{3·6} = 147·03$	$6·6^{3·6} = 891·99$	$11·5^{3·6} = 6584·3$
$4·1^{3·6} = 160·70$	$6·7^{3·6} = 941·61$	$11·75^{3·6} = 7114·4$
$4·2^{3·6} = 175·26$	$6·75^{3·6} = 967·15$	$12·0^{3·6} = 7674·5$

TABLE II.—*Powers of lengths.*

$1^{1·7}$= 1	$9^{1·7}$= 41·900	$17^{1·7}$=123·53
$2^{1·7}$= 3·2490	$10^{1·7}$= 50·119	$18^{1·7}$=136·13
$3^{1·7}$= 6·4730	$11^{1·7}$= 58·934	$19^{1·7}$=149·24
$4^{1·7}$=10·556	$12^{1·7}$= 68·329	$20^{1·7}$=162·84
$5^{1·7}$=15·426	$13^{1·7}$= 78·289	$21^{1·7}$=176·92
$6^{1·7}$=21·031	$14^{1·7}$= 88·801	$22^{1·7}$=191·48
$7^{1·7}$=27·332	$15^{1·7}$= 99·851	$23^{1·7}$=206·51
$8^{1·7}$=34·297	$16^{1·7}$=111·43	$24^{1·7}$=222·00

44. As an example, suppose it were required to find the strength of a hollow cylindrical cast iron pillar, 14 feet long, 6·2 inches external diameter, and 4·1 inches internal; the pillar being flat, and well supported, at the ends.

From the Tables we obtain $14^{1·7}$=88·801, $6·2^{3·6}$ =712·22, and $4·1^{3·6}$=160·70. Whence strength $= 44·3 \times \frac{D^{3·6} - d^{3·6}}{l^{1·7}} = 44·3 \times \frac{712·22 - 160·70}{88·801} = 275·1$ tons.

STRENGTH OF SHORT FLEXIBLE PILLARS.

45. The formulæ above apply to all pillars whose length is not less than about 30 times the external diameter; for pillars shorter than this, it will be necessary to modify the formulæ by other considerations, since in these shorter pillars the breaking weight is a considerable proportion of that necessary to crush the pillar.

46. Thus, considering the pillar as having two functions, one to support the weight, and the other to resist flexure, it follows that when the material is incompressible (supposing such to exist), or when the pressure necessary to break the pillar is very

small, on account of the greatness of its length compared with its lateral dimensions, then the strength of the whole transverse section of the pillar will be employed in resisting flexure; when the breaking pressure is half of what would be required to crush the material, one half only of the strength may be considered as available for resistance to flexure, whilst the other half is employed to resist crushing; and when, through the shortness of the pillar, the breaking weight is so great as to be nearly equal to the crushing force, we may consider that no part of the strength of the pillar is applied to resist flexure.

47. This reasoning is supported by the results from a number of short pillars of various lengths, from 26 times the diameter down to twice the diameter. These pillars were reduced to about half the strength, as calculated by the preceding formulæ, when the length was so small that the breaking weight was half of that which would crush the pillar; and the results from short pillars of other lengths were in accordance with the preceding reasoning. (See Researches on the strength of pillars, &c., art. 39, Phil. Trans. 1840.)

48. We may therefore separate these effects by taking, in imagination, from the pillar (by reducing its breadth, since the strength is as the breadth,) as much as would support the pressure, and considering the remainder as resisting flexure to the degrees indicated by the previous rules.

49. Suppose, then, c to be the force which would crush the pillar without flexure; d the utmost pressure the pillar, as flexible, would bear to break it without being weakened by crushing (as was shown to take place with a certain pressure dependent on the material); b the breaking weight, as calculated by the preceding formulæ for long pillars; y the real breaking weight.

Then supposing a portion of the pillar, equal to what would just be crushed by the pressure d, to be taken away, we have $c - d =$ the crushing strength of the remaining part, and $y - d$ the weight actually laid upon it. Whence $\frac{y-d}{c-d} =$ the part of this remaining portion of the pillar which has to resist crushing,

$$\therefore 1 - \frac{y-d}{c-d} = \frac{c-y}{c-d} =$$

the part to sustain flexure.

50. But the strength of the pillar, if rectangular, may be supposed to be reduced, by reducing either its breadth, or the computed strength of the whole, to the degree indicated by the fraction last obtained. In circular pillars this mode is not strictly applicable; but we obtain a near approximation to the breaking weight y, by reducing the calculated value of b in that proportion.

Whence $b \times \frac{c-y}{c-d} = y$, the strength of a short flexible pillar, b being that of a long one, $\therefore bc - by = cy - dy$,

$$\text{and } y = \frac{b\,c}{b + c - d}.$$

51. It was shown (Experimental Researches, art. 5-7) that cast iron pillars with flat ends uniformly bore about three times as much as those of the same dimensions with rounded ends; and this was found by experiment to apply to all pillars from 121 times the diameter down to 30 times.

52. In flat-ended cast iron pillars, shorter than this, there was observed to be a falling off in the strength; and the same was found to be the case in pillars of other materials, on which many experiments were made, to ascertain whether the results previously mentioned, as obtained from the cast iron pillars, were general. The cause of the shorter pillars falling off in strength, as mentioned above, was rendered very probable by the experiments upon wrought iron; for in that metal a pressure of from 10 to 12 tons per square inch produced a permanent change in, and reduced the length of short cylinders, subjected to it, (art. 60 of Paper above;) and about the same pressure per square inch of section, when required to break by flexure a wrought iron pillar with flat ends, produced a similar falling off in strength to that which was experienced when a weight per square inch, not widely different from this, was required to break a cast iron pillar with flat ends. The fact of cast iron pillars sustaining a marked diminution of their breaking strength by a weight nearly the same as that which produced

incipient crushing in wrought iron, and a falling off in the strength of wrought iron pillars, rendered it extremely probable that the same cause (incipient crushing or derangement of the parts) produced the same change in both these species of iron.

53. The pressure which produced the change mentioned above in the breaking of cast iron pillars was about $\frac{1}{4}$th of that which crushed the material, as given from the experiments upon the metal there used. I shall therefore assume here, as I did there, that one-fourth of the crushing weight is as great a pressure as these cast iron pillars could be loaded with, without their ultimate strength being decreased by incipient crushing; and it was there shown that the length of such a pillar, if solid and with flat ends, would be about 30 times its diameter.

54. We shall have, therefore, $d = \frac{1}{4}$, in the preceding formula

$$y = \frac{b\,c}{b + c - d};$$

whence in cast iron of the kind used, (Low Moor, No. 3,)

$$y = \frac{b\,c}{b + \frac{3\,c}{4}}.$$

55. To find the force necessary to crush a square inch of the iron mentioned above, in order that the value of c, which is that which would crush the whole pillar if inflexible, might be computed, I made (art. 55 of the Paper before referred to) ex-

periments upon it, both upon cylinders and rectangles; and the mean strength from five of the results gave, per square inch, 109,801 lbs. = 49·018 tons.

56. The value of y above is compounded of two quantities, b the strength as obtained from one of the formulæ for long flexible pillars (art. 41 of the present Work), and c the crushing force.

57. The following Table, which gives the dimensions and breaking weights of eleven short solid pillars with flat ends, together with the calculated values of b, c, and y, will show what degree of approximation the calculated strength bears to the real.

Here b, the strength in lbs. (arts. 39 and 41),

$$= 98922 \times \frac{d^{3.55}}{l^{1.7}}.$$

58. *Short solid pillars, flat at the ends, fig.* 26.

Diameter of pillar.	Length of pillar.	Value of b.	Value of c.	Breaking weight.	Calculated breaking weight from the formula $y = \frac{bc}{b + \frac{3c}{4}}$
inches.	ft. inches.	lbs.	lbs.	lbs.	lbs.
·50	1·008 = 12·1	8327	21559	7195	7328
·50	·840 = 10·083	11353	21559	8931	8872
·50	·630 = 7·5625	18515	21559	11255	11508
·50	·315 = 3·7812	60155	21559	17468	16992
·777	1·681 = 20·166	16713	52064	15581	15604
·775	1·260 = 15·125	27005	51797	21059	21241
·785	1·008 = 12·1	41300	53142	24287	27043
·768	·840 = 10·083	52096	50865	25923	29363
·777	·630 = 7·5625	88547	52064	32007	36130
1·022	1·681 = 20·1666	44218	90074	31804	35631
1·000	1·260 = 15·125	66746	86238	40250	43797

59. The next Table (abridged from Table X., art. 36) contains the dimensions of thirteen short hollow cylinders of the same iron, together with their real and calculated breaking weights, for comparison as before. Here, as before, the ends are flat, and the lengths less than 30 times the external diameter.

60. *Hollow uniform cylindrical pillars of Low Moor Iron, No. 3.*

Number of experiments.	Length of pillar.	External diameter.	Internal diameter.	Weight of pillar.		Breaking weight.	Value of b.	Value of c.	Calculated breaking weight from formula $y = \frac{bc}{b + \frac{3c}{4}}$	Remarks.
	feet.	inch.	inch.	lbs.	oz.	lbs.	lbs.	lbs.	lbs.	
1	2·5208	1·26	·767	6	2	33679	38807·6	86178·5	32331	Not perfectly sound.
2	2·5208	1·26	·781	6	1	32867	38274	84310	31790	Core not quite in middle. Thickness of metal on opposite sides, 3 : 4.
3	2·1666	1·25	·768	5	2	35302	48461·7	83882	36501	Air bubbles in casting.
4	2·1666	1·17	·752	4	7	31195	36887	69283·2	28764	Core in centre, T : C : : 43 : 74.
5	1·9166	1·16	·7705	3	9	30383	42633	64844·7	30291	Do., T : C : : 11 : 18.
6	1·6805	1·21	·77	3	9	41751	64599·2	75130·6	40128	Core in centre.
7	1·6805	1·14	·805	2	11	27135	46408	56193	29449	
8	1·4166	1·15	·91	1	11	25511	50927	42636	26191	
9	1·3333	1·15	·92	1	9½	25105	54730	41053	26273	
10	1·2604	1·16	·932	1	8	26729	61304·1	41133·8	27364	Core in centre, T : C : : 52 : 64.
11	1·2604	1·08	·77	1	12	27135	61570·2	49457	30863	
12	1·1667	1·15	·792	2	0	37285	91909	59953	40257	
13	·7333	1·13	·91	0	13½	34037	133000	38704	31750	

Some of these pillars broke into many pieces; several of them were bored inside, and turned on the outside. They were, with the two exceptions named, very good castings. By the ratio T : C is to be understood the depth of the part extended to that compressed in the section of fracture. By the value of b is to be understood the breaking weight, as calculated from the formula

$$b = 99318 \frac{D^{3.55} - d^{3.55}}{l^{1.7}},$$

for the strength of long hollow pillars in lbs., which is given, somewhat abridged, in art. 41.

It will be observed that the calculated strengths agree moderately well with the real ones in both of the preceding Tables, showing that the resistance to crushing is an element of the strength of *short* flexible pillars at least.

COMPARATIVE STRENGTH OF LONG SIMILAR PILLARS.

61. It has been stated (art. 39, results 5 and 6) that the strength of solid pillars with rounded ends varied as $\frac{d^{3.76}}{l^{1.7}}$, and that of those with flat ends as $\frac{d^{3.55}}{l^{1.7}}$ This was when the former pillars were not shorter than about 15, nor the latter than about 30 times the diameter.

62. In the research for the above numbers, I was led to conclude that, if the material had been incompressible, the 3·76 and 3·55 would each have become 4, and the 1·7 have been 2 (see arts. 24 and

33 of the Paper above referred to). In that case the strength would have varied as $\frac{d^4}{l^2}$, which is the ratio of the strength of pillars according to the theory of Euler; which theory was intended to apply to the power of pillars to resist incipient flexure, whilst my inquiry was as to the breaking strength. In similar pillars the length is to the diameter in a constant ratio: calling then the length $n\,d$, where n is a constant quantity, we have, in these different cases, the strength as

$$\frac{d^{3.76}}{n^{1.7} \times d^{1.7}}, \quad \frac{d^{3.55}}{n^{1.7} \times d^{1.7}}, \quad \frac{d^4}{n^2\,d^2}.$$

Dividing, these become

$$\frac{d^{2.06}}{n^{1.7}}, \quad \frac{d^{1.85}}{n^{1.7}}, \quad \frac{d^2}{n^2}.$$

63. In the first of these cases the strength varies as a power of the diameter somewhat higher than the square; in the second somewhat lower; and in the third, as the square. We may therefore conclude, that in similar pillars the strength is nearly as the square of the diameter, or of any other linear dimension; and as the area of the section is as the square of the diameter, the strength is nearly as the area of the transverse section.

64. In deducing the conclusion in the last article, Euler remarks that if, of two similar pillars of the same material, one be double the linear dimensions of the other, the larger will but bear four times as

much as the smaller, though its weight is eight times as great. Berlin Memoirs, 1757.

65. The following Table, containing the results from such of my experiments on solid uniform cylindrical pillars as were from models *similar* in form, will show how far the above conclusions agree with the results of experiments.

	Diameters of pillars compared.	Length of pillars compared.	Breaking weight of pillars.	Powers of the dimensions to which the breaking weights are proportional.	
	inch.	inches.	lbs.		
Pillars with rounded ends.	·497 ·99	7·5625 15·125	5262 19752	1·908	Mean from the powers, 1·865.
	·76 1·52	15·125 30·25	9223 32531	1·819	
	·99 1·97	30·25 60·5	6105 25403	2·057	
Pillars with flat ends.	·51 1·56	20·166 60·5	3830 28962	1·841	
	·50 ·997	30·25 60·5	1662 6238	1·9081	
	·51 1·02	15·125 30·25	6764 21844	1·6913	
	·50 1·022	10·083 20·166	8931 31804	1·8323	

66. In the preceding Table, the pillars being from similar models were assumed to be similar, notwithstanding slight deviations in the measures. It appears that the power of the lineal dimensions,

according to which their strengths vary, is somewhat lower than the second.

67. If long pillars be so formed as to resist being crushed by the breaking weight, as has been mentioned before, they will be similar.

We have seen (art. 52-3) that when pillars require a force to break them by flexure, which exceeds a certain portion of the force which would crush them, if they were not flexible, the pillar sustains a considerable diminution in its power of resistance to flexure in consequence of a partial crushing, or crippling of the material. Suppose $c\,d^2 =$ the crushing force of the pillar (d being the diameter), or that pressure which would cause rupture in it, if it were too short to break by flexure; and $n\,c\,d^2$ that part of this pressure which is the utmost it would, as flexible, sustain without apparent crippling or crushing. Then, since the strength in lbs. to resist fracture by flexure in pillars, with both ends rounded, and both flat, was $33379\,\frac{d^{3.76}}{l^{1.7}}$, and $98922\,\frac{d^{3.55}}{l^{1.7}}$, respectively, as appeared from my experiments, l being in feet and d in inches, we have these quantities each equal to $n\,c\,d^2$, in the cases where short pillars, which break by flexure, are bearing, at the time of fracture, the greatest weights they can sustain without any apparent crushing. Whence, in pillars with rounded ends,

$$33379 \frac{d^{3.76}}{l^{1.7}} = n\,c\,d^2 ; \therefore l = \left(\frac{33379}{n\,c}\right)^{\frac{1}{1\cdot7}} \times d^{\frac{1\cdot76}{1\cdot7}} ;$$

in pillars with flat ends,

$$98922 \frac{d^{3.55}}{l^{1.7}} = n\,c\,d^2 ; \therefore l = \left(\frac{98922}{n\,c}\right)^{\frac{1}{1\cdot7}} \times d^{\frac{1\cdot55}{1\cdot7}}$$

68. In the former of these cases, l varies somewhat faster than as the first power of the diameter, and in the second somewhat slower; the two showing that, in the case of pillars equally loaded to resist crushing by the weight, the length to the diameter will be nearly in a constant ratio, or the pillars must be similar.

ON THE STRENGTH OF PILLARS OF VARIOUS FORMS, AND DIFFERENT MODES OF FIXING.

69. In hollow pillars of greater diameter at one end than the other, or in the middle than at the ends, as in Table XI. (art. 36), it was not found that any additional strength was obtained over that of uniform cylindrical pillars; on the other hand, the strength of these seemed to be the greater; with respect to this, however, the conclusions were not very decisive. The result from the comparison is in agreement with what may be deduced from Euler's theoretical values of the strengths of uniform cylindrical solid pillars, and of those in the shape of a truncated cone (Berlin Memoirs, 1757); his formulæ for these being

$$P = \frac{a^2\,D^4}{A^2\,d^4} \cdot p, \text{ and } P' = \frac{a^2\,D'^2\,E^2}{A^2\,d^4} \cdot p.$$

These values are to one another, *cæteris paribus*, as D^4 to $D'^2 E^2$; where D is the diameter of the uniform pillar, and D' E the diameters of the two ends of that in the form of a truncated cone. But if we compute the diameter of an uniform cylindrical pillar of the same length and solid content as one with unequal diameters, we shall find the uniform pillar stronger than the other, and the more so according as the inequality of the diameters of the latter is greater.

70. The strength of a pillar in the form of the connecting rod of a steam engine was found to be very small; indeed, less than half the strength that the same metal would have given if cast in the form of an uniform hollow cylinder. The ratio of the strength, according to my experiments, was 17578 to 39645.

71. A pillar irregularly fixed, so that the pressure would be in the direction of the diagonal, is reduced to one-third of its strength, the case being nearly similar to that of a pillar with rounded ends, the strength of which has been shown to be only $\frac{1}{3}$rd of that of a pillar with flat ends.[9]

[9] Tredgold, art. 283 of his Work on Cast Iron, and in his Treatise on Carpentry, following the idea of Serlio in his Architecture, recommends circular abutting joints, to lessen the effect of irregularity in the strains upon columns, from settlements and other causes; but this, we see, is voluntarily throwing away two-thirds of the full strength of the material to prevent what may often be avoided.

72. Uniform pillars fixed at one end, and moveable at the other, as in those flat at one end and rounded at the other, break at $\frac{1}{3}$rd of the length (nearly) from the moveable end; therefore, to economize the metal, they should be rendered stronger there than in other parts.

73. Of rectangular pillars of timber it was proved experimentally that the pillar of greatest strength, where the length and quantity of material is the same, is a square.

COMPARATIVE STRENGTHS OF LONG PILLARS OF CAST IRON, WROUGHT IRON, STEEL, AND TIMBER.

74. It results from the experiments upon pillars of the same dimensions, but different materials, that if we call the strength of cast iron 1000, we shall have for wrought iron 1745, cast steel 2518, Dantzic oak 108·8, red deal 78·5. The numbers, all but the last, were obtained from the pillars with rounded ends, and the computations made by the rules used for cast iron.

POWER OF PILLARS TO SUSTAIN LONG CONTINUED PRESSURE.

75. In all the experiments of which an account has been given, the pillars were broken without any regard to time, and an experiment seldom lasted longer than from one to three hours. To determine, therefore, the effect of a load lying constantly upon a pillar, Mr. Fairbairn had at my

suggestion four pillars cast of the same iron as before, and all of the same length and diameter; the length of each was 6 feet, and the diameter 1 inch, and they were rounded at the ends. The first was loaded with 4 cwt., the second with 7 cwt., the third with 10 cwt., and the fourth with 13 cwt.; this last was loaded with $\frac{97}{100}$ of what had previously broken a pillar of the same dimensions, when the weight was carefully laid on without loss of time. The pillar loaded with 13 cwt. bore the weight between five and six months and then broke; that loaded with 10 cwt. is increasing slightly in flexure; the others, though a little bent, do not alter. They have now borne the loads three years. The deflexion of the first pillar is ·01 inch, that of the second 025, and of the third ·409. The deflexion of this last pillar, when first taken, was ·230; and after each succeeding year it was 380, ·380, and ·409, as at present.

EULER'S THEORY OF THE STRENGTH OF PILLARS.

76. It appeared from the researches of this great analyst, that a pillar of any given dimensions and description of material required a certain weight to bend it, even in the slightest degree; and with less than this weight it would not be bent at all (Acad. de Berlin, 1757). Lagrange, in an elaborate essay in the same work, arrives at the same conclusion. The theory as deduced from this conclusion is very beautiful, and Poisson's exposition of

it, in his 'Mécanique,' 2nd edition, vol. i., will well repay the labour of a perusal.

77. I have many times sought, experimentally, with great care for the weight producing incipient flexure, according to the theory of Euler, but have hitherto been unsuccessful. So far as I can see, flexure commences with weights far below those with which pillars are usually loaded in practice. It seems to be produced by weights much smaller than are sufficient to render it capable of being measured. I am therefore doubtful whether such a fixed point will ever be obtained, if indeed it exist. With respect to the conclusions of some writers, that flexure does not take place with less than about half the breaking weight; this, I conceive, could only mean large and palpable flexure; and it is not improbable that the writers were in some degree deceived from their having generally used specimens thicker, compared with their length, than have been usually employed in the present effort.

Some results of the theory of Euler, as given by Poisson (Mécanique, vol. i. 2nd edit.), have been of great service in the course of the inquiry.

78. I will now give the leading results, abridged from four of the Tables of experiments on cast iron pillars, enumerated in art. 36, p. 329; and the reader who wishes for further information upon them, or upon those of wrought iron or timber, is respectfully referred to the original paper in the Philosophical Transactions of the Royal Society, Part II. 1840.

RESULTS OF EXPERIMENTS ON THE RESISTANCE OF SOLID UNIFORM CYLINDERS OF CAST IRON TO A FORCE OF COMPRESSION.

TABLE I.—*Low Moor Iron, No. 3, cast in dry sand. Ends of specimen turned* (fig. 23) *so that the force would pass through the axis.*

Length.	Diameter.	Mean diameter.	Deflexion of middle of pillar.	Corresponding weight.	Breaking weight.	Mean from breaking weights.	Ratio T : C.	Remarks.
inches.	inch.	inch.	inch.	lbs.	lbs.	lbs.		
60·5	·50	....	·07	58				These pillars, and those of the same length below, were made by mistake ½ an inch longer than was intended, and therefore all the future ones were made of the same length, or exact subdivisions of it.
		·50	·49	113	136			
60·5	·50	....	·04	97		143		
			·23	136	150			
60·5	·77	77	....	....	780	780		
60·5	·77				780			
60·5	·99		·05	515				
			·14	991				
		·99	·19	1183				
			·52	1615	1663	1902		
60·5	·99	....	....	....	2141			

60·5	1·28	1·29	·25	5069	5293	5707		
60·5	1·30		·07	5673				
			·10 ?	5897	6121			
60·5	1·29		·03	2141				The two following pillars were cast in green sand.
		1·295	·17	4549				
			·34	4997	5149	. . . .	. . . .	Weight of pillar 19 ℔s. 8 oz.
60·5	1·30		·00 ?	2141				
			bent.	2757		5465		
			·07	5445	5781	. . . .	. . .	19 ℔s. 14 oz.
60·5	1·53	1·52	·14	10525	10861	10861	126 : 26	
60·5	1·51				10861			
60·5	1·53				10121	. . . .	. . . .	28 ℔s 5 oz. cast in green sand.
		1·535				10650		
60·5	1·54				11179	. . . .	. . . .	28 ℔s 9 oz. cast in green sand.
60·5	1·76		·01 ?	8483				
		1·765	·09	11035				
			·30	14225	14701	15560		
60·5	1·77	. . . .	. . . .	. . . .	16419			
60·5	1·76		bent.	2808				
			·02	8617				
		1·78	·12	13721				
			·22	15233	16493			Weight of pillar 38 ℔s.
60·5	1·80		bent.	3355				
			·09	14201		17564		
			·48	18355	18635	. . . .	145 : 36	

TABLE I.—*Continued.*

Length.	Diameter.	Mean diameter.	Deflexion of middle of pillar.	Corres-ponding weight.	Breaking weight.	Mean from breaking weights.	Ratio T : C.	Remarks.
inches.	inch.	inch.	inch.	lbs.	lbs.	lbs.		
60·5	1·94	....	....	....	22811	22811		44 lbs. = weight of pillar.
60·5	1·97 }		bent.	3355				
			·07	12970				
			·28	22127				
		1·96	·50 }	24311				
			·52 }					
			·60	24857	25403 }			47 lbs. ,, ,,
60·5	1·96 }		·03	12287		24291		
			·19	19943				
			·48	22787	23179 }		153 : 43 or 150 : 42	46 lbs. 8 oz. ,, ,,
30·25	·50 }		bent.	248				
			·15	472	526 }			
30·25	·50 }	·50	·02	304		539		
			·09	472	535 }			
30·25	·50 }		....	....	556 }			
30·25	·77	..	·02	1717				
		·77	·10	2390	2726	2726		
30·25	·99 }		·04	2745				
		·99	·13	4985	6105 }		76 : 23	
30·25	·99 }		·02	3641		6105		
			·07	4985	6105 }		78 : 21	This bent apparently in different directions.

30·25	1·29		·01	12287				
		1·29	·07	16115	17515		96 : 34	
30·25	1·29		·08	12287		17235		
			·21	16115	16955		105 : 24 ?	Small flaw in tensile part.
30·25	1·52	....	·04	22619				
			·12	30739	32419		99 : 51	
30·25	1·53	1·52	....	....	34638	32531	101 : 52	
30·25	1·51	....	·07	22619	30536		110 : 43	
20·1666	1·00	1·01	....	....	15737	15737	63 : 67	4 lbs. 3 oz. = weight of pillar.
20·1666	1·02		....	....	15737		65 : 37	4 ,, 5 ,, ,, ,,
20·1666	·785	·767	....	....	7255	6602		2 lbs. 8½ oz. = weight of pillar.
20·1666	·75		....	....	5950			2 ,, 6 ,, ,, ,,
15·125	·50	....	·05	1689				
			·31	1997	1997		39 : 11	
15·125	·50	....	·20	1801				
			·34	1857	1857		39 : 11	
15·125	·50	....	·08	1353				
		·50	·20	1801	1857	1904		
15·125	·77	....	....	....	10138		48 : 29	
15·125	·76	....	....	....	9746		53 : 23	
15·125	·75	·76	....	....	7786	9223	49 : 26	
15·125	·99	....	....	....	20163		57 : 42	
15·125	·99	....	....	....	19239		60 : 39	
15·125	·99	·99	....	....	19855	19752	60 : 39	

TABLE I.—*Continued.*

Length.	Diameter.	Mean diameter.	Deflexion of middle of pillar.	Corresponding weight.	Breaking weight.	Mean from breaking weight.	Ratio T : C.	Remarks.
inches.	inch.	inch.	inch.	lbs.	lbs.	lbs.		
10·083	·76		....	....	16683			The first and second failed by the ends becoming split by a conical wedge which formed at them.
10·083	·76	·76	....	....	16683	17506		
10·083	·77		....	....	19152			
7·5625	·51	....	....	....	6188		31 : 20	
7·5625	·49	....	....	....	4578		32 : 17	
7·5625	·49	·497	....	....	5019	5262	34 : 15	
7·5625	·77	·77	....	....	23893		35 : 42	These pillars were split at both ends.
7·5625	·77		....	....	22003	22948	41 : 36	
3·7812	·50	....	....	....	15233	15107		In these two experiments the area compressed seemed greater than the extended area. The ends were split by the pressure.
3·7812	·50	·50	....	....	14981			

By the letters T C are to be understood the versed sines, or depths of the neutral line, on the surfaces submitted to tension or compression; and the ratio T : C is that of those versed sines or depths, as nearly as the curve of the neutral line could be represented by a straight line. In almost every case the pillars broke nearly in the middle, both in this Table and the following one.

TABLE II.—*Low Moor Iron, No. 3, cast in green sand. Ends of cylinders turned* flat, *and parallel to each other, and the pressure caused by the approach of two parallel surfaces, between which the cylinder was placed, its ends perfectly coinciding with them,* fig. 26.

Length.	Dia-meter.	Mean dia-meter.	Weight.		Deflexion of middle of pillar.	Corres-ponding weight.	Breaking weight.	Mean from breaking weight.	Ratio T : C.	Remarks.
inches.	inch.	inch.	lbs.	oz.	inch.	lbs.	lbs.	lbs.		
60·5	·51	·51	3	$4\frac{1}{4}$	....	....	483	487		These two had disks 2 inches diameter upon the ends: all the rest had the ends turned flat, and were without disks.
60·5	·51		3	$5\frac{1}{4}$			491			
60·5	·77	·77			·07	1162				
					·22	2036				
60·5	·77		7	$5\frac{1}{2}$	·30	2260	2316			
					·06	1588		2456		
					·10	2036				
					·17	2484	2596			
60·5	·99				·05	4123				
					·30	6475	6811			
60·5	1·01	·997			·05	4123				
					·14	5467	5971	6238		
60·5	·99	....	11	8	·04	4123				
					·10	5467	5932			

TABLE II.—*Continued.*

Length.	Diameter.	Mean diameter.	Weight.		Deflexion of middle of pillar.	Corresponding weight.	Breaking weight.	Mean from breaking weight.	Ratio T : C.	Remarks.
inches.	inch.	inch.	lbs.	oz.	inch.	lbs.	lbs.	lbs.		
60·5	1·30	....	....		·10	11235				
					·24	14763				
					·47	16331	16527			
60·5	1·29	1·29	19	11	·15	11217		16064		
					·45	15137	15333			
60·5	1·28	....	20	0	....	....	16331			
60·5	1·55		28	10	·13	21857				
					·21	25105				
					·62	27135	27135			
60·5	1·57		29	7	·10	21857				
		1·56			·14	25105				
					·26	29977		28962		
					·48	31398				
							31398		120 : 37	A wedge broke out, and showed the neutral line.
60·5	1·55		28	5	....	....	28353			
30·25	·50	·50	1	10½	·05	1090				
					·13	1606				
							1662	1662		

30·25	·78			·08	4357				
				·22	7717	8389			
30·25	·78			·04	2905				
		·77		·06	4697		8811		
				·10	8281				
				·19	9177	9625			
30·25	·76			....	....	8420		61 : 15	A crack showed the neutral line.
30·25	1·01			....	....	19132			
30·25	1·00			·05	13833				
		1·01		·19	17865	18369			
30·25	1·02			·05	17795		20310		
				·16	21715	21844		64 : 39	Cracked at neutral line.
30·25	1·00			·05	14841				
				·07	17865				
				·13	20889	21897		65 : 35	

TABLE II.—*Continued.*

Length.	Diameter.	Mean dia-meter.	Weight.	Mean weight.	Breaking weight.	Mean breaking weight.	Ratio T : C.	Remarks.
inches. 20·1666 20·1666	inch. ·51 ·51	inch. ·51	lbs. oz. 1 $1\frac{1}{2}$ 1 $1\frac{1}{4}$	lbs. oz. 1 $1\frac{3}{8}$	lbs. 3830 3830	lbs. 3830		
20·1666 20·1666 20·1666	·78 ·78 ·77	·777	2 9 2 $8\frac{1}{2}$ 2 $7\frac{1}{2}$	2 $8\frac{1}{3}$	16701 15357 14685	15581	45 : 33	The first broke in two pieces near to the middle, and near to one end; and a piece was split off the other end, at the neutral line. The other two broke nearly in the same manner.
20·1666 20·1666	1·03 1·015	1·022	4 6 4 5	4 $5\frac{1}{2}$	32007 31601	31804	56 : 46 54 : 49	Broke in two places near middle; both ends cracked at neutral line.
15·125 15·125 15·125	·51 ·51 ·51	·51	14 $13\frac{3}{4}$ $13\frac{3}{4}$	$13\frac{10}{12}$	6512 7016 6764	6764	32 : 19	
15·125 15·125 15·125 15·125	·79 ·77 ·76 ·78	·775	$29\frac{3}{4}$ 29 $28\frac{1}{2}$ $29\frac{1}{2}$	$29\frac{3}{16}$	21179 22363 19003 21691	21509		
15·125 15·125	1·00 1·00	1·00	3 4 3 4	3 4	39112 41388	40250		

12·1 12·1 12·1	·50 ·50 ·50	·50	$10\frac{1}{2}$ 10 10	$10\frac{1}{6}$	7279 6943 7363	7195	3 : 2 3 : 2 3 : 2	
12·1 12·1 12·1 12·1	·78 ·79 ·79 ·78	·785	1 $7\frac{1}{2}$ 1 8 1 $8\frac{1}{4}$ 1 $7\frac{1}{2}$	1 $7\frac{3}{4}$	25355 24043 23875 23875	24287	39 : 39 1 : 1 1 : 1	
10·0833 10·0833 10·0833	·50 ·50 ·50	·50	$8\frac{1}{2}$ $8\frac{1}{2}$ $8\frac{1}{2}$	$8\frac{1}{2}$	8287 8623 9883	8931	13 : 12 13 : 12 13 : 13	
10·0833 10·0833 10·0833 10·0833	·78 ·77 ·76 ·76	·768			27491 25531 25531 25139	25923	1 : 1	These generally broke in several pieces; but always in the middle by bending. There was, however, usually a wedge formed about the centre, which tended to split the pillar there.
7·5625 7·5625 7·5625	·50 ·50 ·50	·50	$6\frac{1}{3}$ $6\frac{1}{3}$ $6\frac{1}{3}$	$6\frac{1}{3}$	11479 11143 11143	11255	12 : 13 12 : 13 12 : 13	
7·5625 7·5625 7·5625	·78 ·78 ·77	·777	$14\frac{1}{2}$ $14\frac{1}{2}$ $14\frac{1}{4}$	$14\frac{5}{12}$	33225 31601 31195	32007	4 : 5	There was a good deal of doubt respecting the neutral line; but somewhat more than one-half was compressed.
3·7812 3·7812 3·7812	·50 ·50 ·50	·50			17795 17935 16675	17468	20 : 30	These broke in the middle by bending, as before; but they generally showed a short ridge or wedge in the centre, as mentioned above.

TABLE II.—*Continued.*

Length.	Diameter.	Mean diameter.	Weight.	Mean weight.	Breaking weight.	Mean breaking weight.	Ratio T : C.	Remarks.
inches.	inch.	inch.	lbs. oz.	lbs. oz.	lbs.	lbs.		
2 2	·52 ·52	·52			23035 22699	22867		The first bent, and slid off in A B. The other bent and cracked half across in the middle.
1 1 1	·52 ·52 ·52	·52			23963 24747 25139	24616		Broke by a wedge, about three-quarters of an inch high, sliding off in the direction A B.

TABLE III.—*Hollow cylindrical pillars, rounded at the ends* (Pl. II. fig. 36).—*Results of experiments on the strength of hollow uniform cylinders of cast iron (the Low Moor, No. 3), the ends having hemispherical caps* (a) *on them, that the compressing force might act through the axis of the pillar, and its ends move freely. Length of cylinder, including caps on the ends, 7 feet 6¾ inches. The pillars were in most cases, except otherwise mentioned, cast in dry sand, both in this Table and the following one. The weights of the cylinders, as set down, are for the whole length, 7 feet 6¾ inches.*

Number of experiment.	Description of pillar.	Deflexion.	Weight producing the deflexion.	Breaking weight, or that with which the pillar sunk.	Value of x from formula $x=\frac{W}{D^{3.76}-d^{3.76}}$ where W = the breaking weight, and D, d the external and internal diameters.	Remarks.
		inch.	lbs.	lbs.	lbs.	
1	Hollow uniform cylinder. External diameter 1·78 in. Internal do. 1·21 Weight of cylinder 31 lbs.	·03 ·32 ·49	2237 4829 5333	5585	834·37	With the weight 5585 lbs. the pillar sunk down, but was not allowed to bend so as to break. Its elasticity was very little injured, and it was experimented upon in another way, without showing any defect of strength, as was the case with other pillars treated in the same manner.
2	External diameter 1·74 in. Internal do. 1·187 Weight of cylinder 30¼ lbs.	·02 ·13 ·48	2141 4325 5585	5711	933·13	This pillar sunk with the weight 5711, but it was bent no farther than necessary, and was preserved for another experiment, as before.
3	External diameter 2·01 in. Internal do. 1·415 Weight of cylinder 36½ lbs.	·04 ·31 ·75	2237 6845 8105	8357	826·20	The thickness of the metal at the place of fracture varied on the opposite sides as 19 to 42.

TABLE III.—*Continued.*

Number of experiment.	Description of pillar.	Deflexion.	Weight producing the deflexion.	Breaking weight, or that with which the pillar sunk.	Value of x from formula $x=\frac{W}{D^{3\cdot76}-d^{3\cdot76}}$ where W = the breaking weight, and D, d the external and internal diameters.	Remarks.
		inch.	℔s.	℔s.	℔s.	
4	External diameter 2·33 in. Internal do. 1·70 Weight of cylinder 46½ ℔s.	·24 ·37 ·72	11169 12737 14697	15089	903·28	Thickness of metal on opposite sides at place of fracture as 1 to 4 nearly. The thin side was that which was compressed, and the same was the case in most of the other pillars.
5	External diameter 2·23 in. Internal do. 1·54 Weight of cylinder 47 ℔s.	·01 ·22 ·69	2237 8357 12137	12389	808·21	
6	External diameter 2·24 in. Internal do. 1·735 Weight of cylinder 34¾ ℔s.	·015 ·34 ·38	2141 12445 12669	13341	1041·7	
7	External diameter 2·24 in. Internal do. 1·58	·02 ·21	4325 13521	13913	917·62	This pillar was reduced to half its thickness near to the ends, and to three-fourths half-way between the middle and each end, but it did not fail in the reduced parts: it sunk by flexure.
8	External diameter 2·49 in. Internal do. 1·89 Weight of cylinder 48¾ ℔s.	·01 ? ·40 ·52	4123 18623 19239	19855	996·24	This pillar was reduced in the same manner as the last, and sunk by flexure, as before.

9	External diameter 2·47 in. Internal do. 1·98 Weight of cylinder 41 lbs.	·01 ? ·23 ·62	3211 17391 18667	19003	1123·5	The thickness of metal at place of fracture varied as 7 to 9.
10	External diameter 2·46 in. Internal do. 1·855 Weight of cylinder 49 lbs.	bent. ·04 ·49 ·65	2141 9103 18083 18615	19147	989·95	The metal in this varied in thickness at the place of fracture as 3 to 4.
11	External diameter 2·73 in. Internal do. 2·17 Weight of cylinder 48 lbs.	bent. ·03 ·70	3603 12105 22787	23963	949·48	
12	External diameter 2·74 in. Internal do. 2·155 Weight of cylinder 51¼ lbs.	·02 ·14 1·10	3603 21219 27491	27883	1059·5	Variation of thickness of metal at place of fracture 2 to 3 nearly.
13	External diameter 3·01 in. Internal do. 2·48 Weight of cylinder 50¼ lbs.	·07 ·23 ·75	16115 21219 25923	26707	819·46	The thickness of metal at the place of fracture varied in this as 9 to 15.
14	External diameter 3·36 in. Internal do. 2·823 Weight of cylinder 59½ lbs.	·09 ·32 1·10	16115 30627 40335	40973	895·02	Variation of thickness of metal at the place of fracture 19 to 34.
15	External diameter 3·36 in. Internal do. 2·63 Weight of cylinder 77¾ lbs.	bent. ·09 ·30 ·90 1·07	3355 16115 33824 48511 49494	50477	880·11	Variation of metal in thickness at point of fracture 5 to 7.

TABLE III.—*Continued.*

Number of experiment.	Description of pillar.	Deflexion.	Weight producing the deflexion.	Breaking weight, or that with which the pillar sunk.	Value of x from formula $x=\frac{W}{D^{3\cdot76}-d^{3\cdot76}}$ where W = the breaking weight, and D, d the external and internal diameters.	Remarks.
		inch.	lbs.	lbs.	lbs.	
16	Solid uniform pillar, rounded at the ends	bent.	2141			
	Diameter 2·24 in.	·02	7541			
	Weight 93 lbs.	·25	20273	21281	1026·6	
17	Pillars of shorter lengths than those above. External diameter 1·78 in. Internal do. 1·21 Length 4 feet 9 inches.			13693	927·86	
18	External diameter 2·31 in. Internal do. 1·67 Length 4 feet 9 inches.			36382	1005·3	Thickness of metal on opposite sides 10 to 21.
19	External diameter 1·85 in. Internal do. 1·36 Length 2 feet 7 inches. Weight of 2 feet 5 inches =8 lbs. 15½ oz.	·35	32587	33763	784·9	The metal in this varied in thickness at point of fracture as 11 to 15. The great weight necessary to break this very short pillar probably caused incipient crushing, and thus reduced the value of x.

The value of x, obtained in the sixth column of the preceding Table, is the strength of a solid pillar, 1 inch diameter, and of the same length as those above. For, when the length is constant, the strength W varies as $D^{3\cdot76}-d^{3\cdot76}$, (arts. 39 & 41); and therefore, to find x, the strength of a solid pillar, 1 inch diameter and 7 feet 6¾ inches long, we have $D^{3\cdot76}-d^{3\cdot76} : 1^{3\cdot76} :: W : x = \frac{W}{D^{3\cdot76}-d^{3\cdot76}}$. The mean from the values of x found in the Table above is 932·76 lbs., and if this be multiplied by $l^{1\cdot7}$, where $l = 7\cdot5625$ feet (= 7 feet 6¾ inches), we obtain the strength in pounds of a solid pillar, 1 foot long and 1 inch diameter, rounded at the ends; and this, divided by 2240, to reduce it into tons, is the co-efficient 12·979, called 13, in the formula for the strength, $13 \times \frac{D^{3\cdot6}-d^{3\cdot6}}{l^{1\cdot7}}$, (art. 41), where the index 3·6 is put as a mean between the 3·76 of pillars with rounded ends, and the 3·55 of those with flat ends. The remarks here made will apply to the values of x in the next Table.

Table IV.—*Hollow Cylindrical Pillars, flat at the ends* (fig. 37). *Results of Experiments on the Strength of Hollow Uniform Cylinders of Cast Iron* (*Low Moor, No.* 3), *the ends being turned flat and perpendicular to the sides, and the pressure communicated by the approach of parallel surfaces, against which the ends of the pillars were firmly bedded. Length of each pillar 7 feet* $6\frac{3}{4}$ *inches, except otherwise mentioned.*

Number of experiment.	Description of pillar.	Deflexion.	Weight producing the deflexion.	Breaking weight, or that with which the pillar sunk.	Value of x from formula $x=\frac{W}{D^{3.55}-d^{3.55}}$ where W = the breaking weight, D, d the external and internal diameters.	Ratio of the thicknesses of the ring of metal on opposite sides at place of fracture.	Remarks.
		inch.	lbs.	lbs.	lbs.		
1	Hollow uniform cylinder, same as in Experiment 1 of the preceding Table. External diameter 1·78 Internal do. 1·21 Length of cylinder 7 ft. $4\frac{3}{4}$ in. Weight (length 7 ft. $6\frac{3}{4}$ in.) 31 lbs.	·02 ·03 ·00 ·05 ·20 ·50 ·66	2813 3821 unloaded 4829 12001 16705 17489	17840	2973·7	1 : 5	When this cylinder was broken, it was found that the thinner side was that which was compressed. The weight of this cylinder and all those below, whether their lengths are 7 feet $6\frac{3}{4}$ inches, or 7 feet $4\frac{3}{4}$ inches, are given for the greater length.
2	Cylinder same as No. 2 in preceding Table External diameter 1·74 Internal do. 1·187 Length of cylinder 7 ft. $4\frac{3}{4}$ in. Weight of cylinder considered uniform as before reduction, $30\frac{1}{4}$ lbs.	·12 ·32 ·48 direction changed ·54	11217 15137 15921 16313	16705	3031·5	7 : 11	This cylinder was reduced to half its thickness near to the ends, and to three-fourths of its thickness half-way between the middle and the ends, but it did not break at the reduced parts.

3	External diameter 1·76 Internal do. 1·18	bent ·03 ·09 ·36 ·54	2141 2749 6677 15233 16241	16745	2968·7	1 : 3	This cylinder was reduced in the same manner as the last, or somewhat more, and the fracture took place at the reduced part, half-way between the middle and one end; the reduced part, as appeared by measure, was a little less than three-fourths of the whole before reduction.
4	External diameter 1·75 Internal do. 1·11 Weight of cylinder 32 lbs.	·01 ·19 ·55	2237 16477 20509	20957	3586·9	1 : 2	This pillar was a good sound casting, and was not reduced in its thickness in the manner of the last two.
5	External diameter 2·04 Internal do. 1·46 Length 7 ft. 4¾ in. Weight 35½ lbs.	·04 ·08 ·37 ·52	3589 14703 29977 31601	32413	3573·8	1 : 1	This column was not reduced in thickness, as in the second and third experiments.
6	External diameter 2·01 Internal do. 1·368 Length of cylinder 7 ft. 6¾ in. Weight 37¾ lbs.	·03 ·08 ·00 ·38 ·53	3589 18667 unloaded 28353 29977	30789	3290·3	7 : 10	Slight bubble in place of fracture. The ends of the cylinder were not reduced.
7	External diameter 2·01 Internal do. 1·415 Length 7 ft. 4¾ in. Weight 36½ lbs.	bent ·10 ·01 ·14 ·25	4251 21857 unloaded 25917 27541	28353	3214·8	5 : 11	This cylinder was the same as that in Experiment 3 of the last Table. It was rendered quite straight, and its ends were firmly bedded; it was reduced, as in Experiment 2, and it broke in the middle, and at three of the reduced places.
8	External diameter 1·99 Internal do. 1·31 Length 7 ft. 5·8 in. Weight before reduct^{n}. 39 lbs.	bent ·20 ·55 ·90	1456 15605 24205 26731	27067	2988·3	6 : 7	This cylinder was reduced in the manner of the preceding ones. It broke at a small flaw near the middle.

TABLE IV.—*Continued.*

Number of experiment.	Description of pillar.	Deflexion.	Weight producing the deflexion.	Breaking weight, or that with which the pillar sunk.	Value of x from formula $x=\frac{W}{D^{3.55}-d^{3.55}}$ where W = the breaking weight, D, d the external and internal diameters.	Ratio of the thicknesses of the ring of metal on opposite sides at place of fracture.	Remarks.
		inch.	lbs.	lbs.	lbs.		
9	External diameter 2·23 Internal do. 1·54 Length 7 ft. 4¾ in. Weight before reduction 47 lbs.			40569	3099·0	4 : 9	This cylinder was the same as that in No. 5, last Table; it was now reduced in thickness in the same manner, and to the same degree, as in the preceding ones. It broke in the middle, and at one of the reduced places near to the middle.
10	Uniform solid cylinder, cast in green (moist) sand Diameter 1·76 Length 7 ft. 6¾ in. Weight 56 lbs.	bent bent ·35 ·65	4135 10855 21219 22787	23179	3115·5		With 23179 lbs. it became bent more than an inch, and slipped out of its place; it was afterwards rendered straight and replaced; and it would have broken with a less weight.
11	Uniform solid cylinder, cast in dry sand Diameter 1·72 Length 7 ft. 6¾ in. Weight 53 lbs. 8 oz.	·20 ·28 ·44 ·65	16115 18355 20595 21715	21995	3207·7		With the last weight, 21995 lbs., the pillar slipped from its fixings, as the preceding one had done, and when replaced it was broken with a less weight than it had borne before. The weight, in both this case and the last, was so near to what the breaking weight must have been, if fracture had been effected as usual, that I have not hesitated to put down the results as those of fracture.

TRANSVERSE STRENGTH.

79. The transverse strain is that to which cast iron and some other materials are most frequently subjected, and therefore experiments have been oftener made that way than any other. Still, as regards cast iron, whose uses are multiplying every day, the knowledge of the practical man has hitherto been far from equalling his wants; and accordingly various efforts have lately been made, and doubtless will continue to be made, to obtain extra information upon so important a subject.

I will give an account of some of these, the objects of which are as below.

1st. To ascertain what alteration takes place in bars of cast iron subjected to long-continued strains.

2nd. To determine the effects of changes in the temperature of bars upon their strength.

3rd. To inquire into the elasticity and strength of cast iron bars, under ordinary circumstances, the time when the former becomes impaired, and the erroneous conclusions that have been deduced from it.

4th. To find the best forms of beams, and the strength of beams of particular forms.

80. LONG-CONTINUED PRESSURE UPON BARS OR BEAMS.

To ascertain how far cast iron beams might be

trusted with loads permanently laid upon them, Mr. Fairbairn made the following experiments.—(Report on the Strength of Cast Iron, obtained from the Hot and Cold Blasts, vol. vi. of the British Association.)

He took bars, both of cold and hot blast iron (Coed Talon, No. 2), each 5 feet long, and cast from a model 1 inch square; and having loaded them in the middle with different weights, with their ends supported on props 4 feet 6 inches asunder, they were left in this position to determine how long they would sustain the loads without breaking. They bore the weights, with one exception, upwards of five years, with small increase of deflexion, though some of them were loaded nearly to the breaking point. Since that time, however, less care has been taken to protect them from accident, and three others have been found broken. They are carefully examined, and have their deflexions taken occasionally, which are set down in the following Table, which contains the exact dimensions of the bars, with the load upon each. These experiments were undertaken by Mr. Fairbairn at my suggestion, as I was led to conceive, from experiments I had made in a different way upon malleable iron, that time would have little effect in destroying the power of beams to bear a dead weight.

81. *Experiments of H. Fairbairn, Esq. on the Strength of Bars to resist long-continued pressure.*

Date of observation.	Temperature of the air at time of observation.	Experiment 1. Cold blast iron. Depth of bar 1·050. Breadth of bar 1·030.	Experiment 2. Hot blast iron. Depth of bar 1·050. Breadth of bar 1·010.	Experiment 3. Cold blast iron. Depth of bar 1·030. Breadth of bar 1·020.	Experiment 4. Hot blast iron. Depth of bar 1·040. Breadth of bar 1·020.	Experiment 5. Cold blast iron. Depth of bar 1·030. Breadth of bar 1·020.	Experiment 6. Hot blast iron. Depth of bar 1·050. Breadth of bar 1·000.	Experiment 7. Cold blast iron. Depth of bar 1·000. Breadth of bar 1·010.	Experiment 8. Cold blast iron. Depth of bar 1·020. Breadth of bar 1·030.	Experiment 9. Hot blast iron. Depth of bar 1·040. Breadth of bar 1·010.
	Fahrenheit.	Deflexions with a permanent load of 280 lbs. laid upon each.		Deflexions with a permanent load of 336 lbs. laid upon each.		Deflexions with a permanent load of 392 lbs. laid upon each.		Deflexions with a permanent load of 448 lbs. laid upon each.		
1837.										
March 6				1·267		1·684	1·715	1·964	1·410	This broke with 392 lbs.; other hot blast bars were tried, but they were successively broken with 448 lbs.
,, 9	49°	·916	1·043	1·270	1·454	1·694	1·758	2·005	1·413	
,, 11		·930	1·064	1·270	1·461	1·694	1·760	2·005	1·413	
,, 17								2·010	1·413	
April 15	47°	·930	1·078	1·271	1·475	1·716	1·767	2·014	1·422	
May 31	62°	·932	1·082	1·274	1·481	1·725	1·775	Broke after bearing the weight 37 days.	1·424	
Aug. 22	70°	·937	1·086	1·288	1·504	1·737	1·783		1·438	
Nov. 18	45°	·942	1·083	1·286	1·499	1·724	1·773		1·431	
1838.										
Jan. 8	38°	·941	1·086	1·288	1·502	1·722	1·773		1·430	
March 12	51°	·945	1·091	1·298	1·505	1·801	1·784		1·439	
June 23	78°	·963	1·107	1·316	1·538	1·824	1·803		1·457	
1839.										
Feb. 7	54°	·950	1·093	1·293	1·524	1·815	1·784		1·433	
July 5	72°	·959	1·104	1·305	1·533	1·824	1·798		1·446	
Nov. 7	50°	·955	1·102	1·303	1·531	1·824	1·796		1·445	
Dec. 9	39°	·956	1·102	1·303	1·531	1·823	1·796		1·445	
1840.										
Feb. 14	50°	·955	1·104	1·305	1·531	1·824	1·797		1·446	
April 27	63°	·954	1·116	1·309	1·519	1·818	1·802		1·445	
June 6	61°	·951	1·112	1·303	1·520	1·825	1·798		1·445	
Aug. 3	74°	·953	1·115	1·305	1·523	1·826	1·801		1·447	
Sept. 14	55°	1·047*	1·115	1·305	1·613*	1·826	1·802		1·447	
1841.										
Nov. 22	50°	1·045	1·115	1·306	1·620	1·829	1·804		1·449	
1842.										
April 19	58°			1·308	1·620	1·828	1·812		1·449	

* After August 3, 1840, a body seems to have fallen upon the bars of the 1st and 4th Experiment, and this may have increased their deflexions.

82. Looking at the results of these experiments, and the note upon the first and fourth, it appears that the deflexion in each of the beams increased considerably for the first twelve or fifteen months; after which time there has been, usually, a smaller increase in their deflexions, though from four to five years have elapsed. The beam, in Experiment 8, which was loaded nearest to its breaking weight, and which would have been broken by a few additional pounds laid on at first, had not, perhaps, up to the time of its fracture, a greater deflexion than it had three or four years before; and the change in deflexion in Experiment 1, where the load is less than $\frac{2}{3}$rds of the breaking weight, seems to have been almost as great as in any other; rendering it not improbable that the deflexion will, in each beam, go on increasing till it become a certain quantity, beyond which, as in that of Experiment 8, it will increase no longer, but remain stationary. The unfortunate fracture of this last beam, probably through accident, has left this conclusion in doubt.

83. These important experiments show that cast iron may be trusted with permanent loads far greater than has previously been expected; it having been generally admitted that about $\frac{1}{3}$rd of the breaking weight was as far as it was safe to load a beam with in practice. It was conceived that a load greater than this would break the beam or other body in time, since the elasticity was thought to be injured with about this weight; and it was

accounted unsafe to load a body beyond its elastic force, (see arts. 70, 71, Part I. of this volume.)

EFFECTS OF TEMPERATURE UPON THE STRENGTH OF CAST IRON.

84. Mr. Fairbairn gave, in the Report previously referred to, the results of several experiments to determine how far the strength of cast iron bearers is influenced by such changes of temperature as they are occasionally subjected to. He had a number of bars cast, of Coed Talon Iron, Nos. 2 and 3, part of them being of iron made with a heated blast, and part with cold. These bars were from models 1 inch square and 2 feet 6 inches long. They were laid on supports 2 feet 3 inches asunder, and broken by weights hung at the middle. The experiments were made in winter. Some bars were broken in the open air; some when immersed in frozen water or covered with snow; some in melted lead; and others when heated red hot. The results are in the following Table, the first column of which shows the temperature under which the experiments were made; the second and third columns give the breaking weights of the bars in pounds, when reduced by calculation to exactly 1 inch square; and the fourth column gives the ratio of the strengths of the cold and hot blast irons in the two preceding columns.

Temperature, Fahrenheit.	Coed Talon, cold blast.		Coed Talon, hot blast.		Ratio of the strengths of the two irons.
	No. 2 Iron. lbs.		No. 2 Iron. lbs.		
16°			800·3		
26°	851·0		823·1		1000 : 967·2
		mean		mean	
32°	940·7 958·5	949·6	933·4 906·0	919·7	1000 : 977·6
190°	743·1		823·6		1000 : 1108·3
Red in the dark	723·1		829·7		
Perceptibly red in daylight	663·3				
	No. 3 Iron.		No. 3 Iron.		
		mean			
212°	905·0 943·6	924·3	818·4		1000 : 885·4
				mean	
600°	909·3 1157·0	1033·1	834·1 917·5	875·8	1000 : 847·7

85. It would appear from these experiments, though the results are somewhat anomalous, that the strength of cast iron is not reduced when its temperature is raised to 600°, which is nearly that of melting lead; and it does not differ very widely whatever the temperature may be, provided the bar be not heated so as to be red hot.

ON THE STRENGTH OF CAST IRON BARS OR BEAMS UNDER ORDINARY CIRCUMSTANCES,—THE TIME WHEN THE ELASTICITY BECOMES IMPAIRED,—AND THE ERRONEOUS CONCLUSIONS THAT HAVE BEEN DERIVED FROM A MISTAKE AS TO THAT TIME.

86. It is, as has been before observed, to ascertain the resistance of materials to a transverse strain

that the efforts of experimenters have chiefly been directed: one reason for this seems to be the great facility with which bodies can be broken this way comparatively with others, which require large weights, or complex machinery, and often considerable attention to theoretical requirements.

The inquiry into the strengths of hot and cold blast cast iron, which I undertook, in conjunction with Mr. Fairbairn, for the British Association, and whose results were inserted in the sixth volume of their Reports, induced me to make a number of experiments on the transverse strength of bars. In these experiments, most of which were made before Mr. Fairbairn's avocations enabled him to attend to the matter, I had a number of bars cast from models 5 feet long and 1 inch square; and the castings were supported on props 4 feet 6 inches asunder, and broken by weights suspended from the middle. These bars were made long and slender, in order that small variations in the elasticity might be rendered obvious. My earliest experiments upon the Carron Iron, and particularly those upon the Buffery, convinced me that the elasticity of bars is injured much earlier than is generally supposed; and that instead of it remaining perfect till one-third, or upwards, of the breaking weight was laid on, as is generally admitted by writers (Tredgold, art. 70, &c.), it was evident that $\frac{1}{5}$th or less produced in some cases a considerable set, or defect of elasticity; and judging from its slow increase afterwards, I was persuaded that it had not come on by any sudden change, but had existed, though in a less degree, from a very early period. I mentioned the fact, and my convictions, some time afterwards to Mr. Fairbairn, and, after examining the matter with more attention than before, expressed a desire to have bars cast of a greater length than the preceding ones, to render the defect more obvious. I had, therefore, two bars of Carron Iron, No. 2, hot blast, cast from the same model, each 7 feet long; they were uniform throughout, and the form of the section of each was as in the figure. They were laid on supports, 6 feet 6 inches asunder, and broken by weights suspended from the middle; the former with the rib downward, in which experiment the flexure would be, almost wholly, owing to the extension of that rib; and the latter with the rib upward, in which the flexure would be owing to the compression of the rib. In both of these bars, the dimensions of the parallelogram A B was the same, $= 5 \times \cdot 30 = 1 \cdot 50$ square inches, the thickness being ·30 inch. In the former of these cases, C D $= 1 \cdot 55$, and D E (which may represent the uniform thickness of the rib D F) $= \cdot 36$ inch. In the latter casting C D $= 1 \cdot 56$, and D E $= \cdot 365$ inch. The results are as follow.

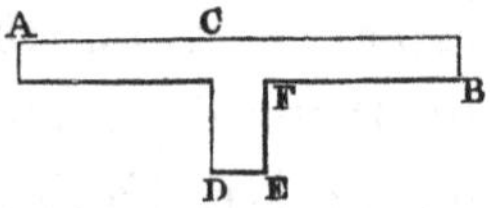

FIRST BAR. Broken with the vertical rib downwards. T			SECOND BAR. Broken with the vertical rib upwards. ⊥		
Weight in lbs.	Deflexion in inches.	Deflexion (load removed.)	Weight in lbs.	Deflexion in inches.	Deflexion (load removed).
7	·015	visible	7	....	not visible
14	·032	·001?	14	·025	visible
21	·046	·002	21	·045	·002
28	·064	·004	28	·065	·003
56	·130	·005	56	·134	·005
112	·273	·020	112	·270	·015
168	·444	·035	224	·580	·058
224	·618	·058	336	·895	·101
280	·813	·093	448	1·224	·155
336	1·030	·130	560	1·585	·235
364	broke	....	672	1·985	·330
			784	2·410	·490
∴ Ultimate deflexion=1·138			896	....	·722
			1008	4·140	1·040
			1064	....	....
			1120	broke	....

During the fracture a wedge 2·92 inches long, and 1·05 deep, broke out, of the form

∴ Ultimate deflexion=5·00

In the first of these experiments, it will be seen that the elasticity was sensibly injured with 7 lbs., and in the latter with 14 lbs.; the breaking weights being 364 lbs. and 1120 lbs. In the former of these cases a set was visible with $\frac{1}{52}$nd, and in the other with $\frac{1}{80}$th of the breaking weight, showing that there is no weight, however small, that will not injure the elasticity. The ratio of the breaking weights in these two experiments was as 1 : 3·07; showing that a bar of this form was more than three times as strong one way up as the opposite way.

87. The mode I used to observe when the elastic force became injured was as follows. When a bar was laid upon the supports for experiment, a "straight edge" was placed over it, the ends of which rested upon the bar directly over the points of support. These ends were slides which enabled the straight edge to be raised or lowered at pleasure. In this manner it was easy to bring it down to touch in the slightest degree a piece of wood tied upon the middle of the bar. A candle was then placed at the side of the bar, opposite to where the observer stood, by the light of which, distances extremely minute could be observed.

88. The results from these experiments will enable us to see the mode by which Tredgold deduced the erroneous conclusion, as to the high tensile strength of cast iron, adverted to in the note to art. 143, page 155; and which has produced an effect on many of the numerical conclusions throughout his work.

The experiment (No. 2, art. 56), from which, and others, he infers that a cast iron bar, 1 inch square and 34 inches between the supports, will bear 300 ℔s. on the middle, without injury to the elasticity, has not, I conceive, been examined in its progress with adequate care to form the basis of important conclusions. It is evident, from my experiments given above, that the elastic force was injured with much less weight than that which formed the set he first noticed in the beam.

Tredgold calculated the direct tensile strength of the cast iron in these experiments, from the results of the transverse strength. He assumed in his calculations, that the position of the neutral line remains fixed, and in the middle of a rectangular or circular section, during the whole experiment; and the resistance of a particle, at equal distances on each side of the neutral line, to be the same.

From these principles he calculated that the most extended surface of the bar was supporting a tension of 15,300 lbs. per square inch, without the elasticity being injured. This number, which is greater than was required to tear asunder the specimen in many of the preceding experiments (art. 3, Part. II.), Tredgold adopts, throughout the work, as the direct tensile strength of cast iron, when not strained beyond the elastic force.

From these principles he calculates (art. 212 of his work) that the greatest extension which cast iron will bear without injury to its elastic force is $\frac{1}{1204}$th part of its length. In art. 70, he calculates the weights which would be required to destroy the elasticity of a number of cast iron bars, the breaking weights of which are severally given. He compares these weights, and concludes that a body requires about three times as much to break it as to destroy its elastic force. Hence he would conclude that the absolute strength, per square inch, of this iron, is $15{,}300 \times 3 = 45{,}900$ lbs. nearly, or more than 20 tons. Tredgold computes, in several cases (arts. 72 to 76), the ultimate tensile strength of cast iron bars from the weights which broke them transversely. He found the strength to vary from 40,000 to 48,200 lbs. per square inch, the mean being 44,620 lbs., or nearly three times 15,300, as mentioned above.

My own experiments, which were made by tearing asunder 25 castings, prepared with great care, from cast iron obtained from various parts of England, Scotland, and Wales, gave as a mean 16,505 lbs. $= 7{\cdot}37$ tons per square inch; and in no case, except one, was the strength found to be more than $8\frac{1}{2}$ tons per square inch (art. 3, page 310).

89. According to the principles assumed by Tredgold (arts. 37 and 100), the position of the neutral line must remain unchanged during flexure by different weights; but experiment shows that the neutral line shifts as the weights are increased; and at the time of fracture it is frequently near to the concave side of the beam. This seldom can be discovered in fractures of beams by simple transverse pressure, but it may sometimes be discovered in those that have been broken by a blow, and then it

will perhaps seem to have been at $\frac{1}{5}$th or $\frac{1}{6}$th of the depth of a rectangular beam, as I have occasionally observed, the smaller part being compressed.

90. From the experiments on the strength of cast iron pillars (Tables I. and II., art. 78) we get additional evidence upon this matter; for in these the neutral line was frequently well defined; and in the longest pillars, those which required the least weight to break them, the ratio T : C, which is that of the depths of the parts submitted to tension and compression, was, in different cases, 126 : 26, 145 : 36, 153 : 43, and 120 : 37. In two rectangular pillars, each $60\frac{1}{2}$ inches long, and $1\frac{1}{2}$ inches square, the area of the compressed part was less than $\frac{1}{7}$th of the whole section. In the pillars mentioned above the weight necessary to break them was very small, compared with that which would have been required to crush them without flexure; and there was scarcely a pillar broken in which the part compressed exceeded the part extended, though some had sustained very great pressures.

In the longest pillars mentioned above we may, I conceive, consider the position of the neutral line as not widely different from what it would have been if the pillar had been broken as a beam, transversely; the only difference in the two cases being, that the compressed part, compared with the extended part, would be greater in the pillar, through the weight laid upon it, than in the beam. And we have seen that, in the pillars alluded to, the depth of the compressed part varied from $\frac{1}{3}$rd to less than $\frac{1}{6}$th of that of the extended part; and in a beam the depth of the compressed part would be still smaller. In the experiments on tension and compression (arts. 33-34, Part II.), it has been shown that cast iron resists crushing with about seven times as much force as it does tearing asunder; and some experiments, not yet published, have inclined me to believe that the neutral line in a rectangular beam, at the time of fracture, divides the section in the proportion of six or seven to one, or one not widely different from this; but I draw the conclusion with considerable diffidence.

91. The subject here treated of will be adverted to in a future article, after the experiments which I am intending to give upon the transverse strength of bars have, with those on the tensile strength (art. 3, page 310), furnished data for the purpose.

EXPERIMENTS TO DETERMINE THE TRANSVERSE STRENGTH OF UNIFORM BARS OF CAST IRON.

92. I will now give the results of a very extensive series of experiments upon rectangular bars, all cast from the same model, and including irons from the principal Works in the United Kingdom.

The great accumulation of specimens of iron, which were obtained, but could not be used in our inquiry, before mentioned, respecting the strength of hot and cold blast iron, afforded a good opportunity to acquire the relative values of many of the leading irons. Mr. Fairbairn, therefore, undertook the matter, and had castings made from the whole; and having increased them by subsequent additions of iron, the variety of results from the whole, both of Mr. Fairbairn's and my own experiments, has become great, especially when those are added to them which we derived from the hot blast inquiry, and these will be found abridged in the following pages. The experiments of Mr. Fairbairn were very carefully made, some of them by myself, as those on the anthracite iron, &c., and all with the same attention to accuracy. The results are published at length in the Manchester Memoirs, vol. vi., second series; but are here given in an abridged form, only one series of results being set down for each kind of iron; every result in each series being a mean between values derived from the same weights in different experiments. The bars were cast from a model, 5 feet long and 1 inch square, and they were, during the experiment, laid upon supports 4 feet 6 inches asunder, and bent by weights suspended from the middle. After each load had been laid on, which was done with care and with small additions of weight, the deflexion was obtained by means of a long scale in the form of a wedge, graduated along its side, so that very minute distances could be measured. The beam was then unloaded in order that the defect of elasticity, or set, might be obtained. These experiments, and those made by my friend for the British Association, were conducted in the same manner as others which I had made some time before on the Carron, Buffery, and other irons, mentioned in art. 86; and the remark which I had made of the very early defect of elasticity of cast iron, as shown by my experiments, received a further confirmation from these; as the quantity of set was, from that time, always carefully observed. The results of all the bars were afterwards reduced by calculation in the same manner as those in my Report (British Association, vol. vi.), in order to preserve uniformity in the whole.

93. *Experiments of W. Fairbairn, Esq., on the strength of uniform rectangular bars of cast iron.*

English Irons.

No. 1. Apedale iron, No. 2, hot blast, Newcastle, Staffordshire.			No. 2. Adelphi iron, No. 2, cold blast, Derbyshire.			No. 3. Butterley iron, Derbyshire.			No. 4. Eagle Foundry iron, No. 2, hot blast, Staffordshire.			No. 5. Level iron, No. 1, hot blast, Staffordshire.			No. 6. Level iron, No. 2, hot blast, Staffordshire.		
Means from 2 experiments.			Means from 2 experiments.			Means from 2 experiments.			Means from 2 experiments.			Means from 2 experiments.			Means from 3 experiments.		
Depth of bar 1·017 in. Breadth „ 1·009 Wt. of 1 of the bars 15 lbs. 3 oz.			Depth of bar 1·030 in. Breadth „ ·993 Weight „ 15 lbs. 8 oz.			Depth of bar ·995 in. Breadth „ ·988 Weight „ 14 lbs. 12½ oz.			Depth of bar 1·024 in. Breadth „ 1·035 Weight „ 15 lbs. 13½ oz.			Depth of bar 1·011 in. Breadth „ 1·010 Weight „ 15 lbs. 6½ oz.			Depth of bar 1·033 in. Breadth „ 1·011 Weight „ 15 lbs. 13 oz.		
Weight in lbs.	Deflexion in inches.	Deflexion, load removed.	Weight in lbs.	Deflexion in inches.	Deflexion, load removed.	Weight in lbs.	Deflexion in inches.	Deflexion, load removed.	Weight in lbs.	Deflexion in inches.	Deflexion, oad removed.	Weight in lbs.	Deflexion in inches.	Deflexion, load removed.	Weight in lbs.	Deflexion in inches.	Deflexion, load removed.
112	·277	·009	30	·065	+	28	·068		14	·032	+	56	·137	+	56	·129	·002
182	·487	·023	56	·135	·002	56	·135	·002	56	·136	·003	112	·273	·012	112	·246	·011
238	·673	·045	112	·294	·014	126	·335	·015	112	·279	·013	168	·429	·025	168	·404	·022
294	·878	·076	168	·470	·034	182	·509	·038	168	·441	·030	224	·598	·044	224	·559	·038
350	1·107	·110	224	·658	·060	238	·702	·064	224	·616	·051	280	·776	·070	280	·726	·061
406	1·356	·162	280	·865	·093	294	·912	·099	280	·803	·078	336	·965	·097	336	·905	·088
441	1·536		336	1·080	·138	350	1·142	·147	336	1·006	·113	392	1·171	·133	392	1·095	·121
455	broke		392	1·335	·201	406	1·402	·203	392	1·225	·159	448	1·392	·185	420	1·196	
			434	1·542		448	1·633		420	1·347		476	broke		453	broke	
			448	1·615		462	1·717		448	broke							
			465	broke		479	broke										
	ult. defl. =1·583			ult. defl. =1·705			ult. defl. =1·823			ult. defl. =1·462			ult. defl. =1·499			ult. defl. =1·314	

English Irons.

No. 7. Low Moor iron, No. 2, cold blast, Yorkshire.			No. 8. Milton iron, No. 1, hot blast, Yorkshire.			No. 9. Milton iron, No. 3, hot blast, Yorkshire.		
Means from 2 experiments.			Means from 2 experiments.			Means from 2 experiments.		
Depth of bar ·999 in. Breadth „ 1·009 Weight „ 14 tbs. 13 oz.			Depth of bar 1·061 in. Breadth „ 1·042 Weight „ 16 tbs. 8½ oz.			Depth of bar 1·023 in. Breadth „ 1·009 Weight „ 15 tbs. 14 oz.		
Weight in tbs.	Deflexion in inches.	Deflexion, load removed.	Weight in tbs.	Deflexion in inches.	Deflexion, load removed.	Weight in tbs.	Deflexion in inches.	Deflexion, load removed.
56	·145	·006	42	·103	+	42	·093	+
112	·301	·013	112	·296	·008	56	·127	+
182	·524	·046	182	·508	·036	126	·294	·011
238	·724	·074	238	·697	·060	182	·442	·028
294	·946	·111	294	·907	·092	238	·599	·045
350	1·195	·163	350	1·143	·137	294	·772	·068
406	1·480	·237	406	1·406		350	·955	·097
455	1·784		413	broke		406	1·156	·137
465	broke					441	1·294	
						451	broke	
	ult. defl. =1·853			ult. defl. =1·437			ult. defl. =1·338	

No. 10. Elsicar iron, No. 2, cold blast.			No. 11. Oldberry iron, No. 2, cold blast.			No. 12. Old Park iron, No. 2, cold blast.		
Means from 2 experiments.			Means from 2 experiments.			Means from 2 experiments.		
Depth of bar 1·025 in. Breadth „ 1·007 Weight „ 15 tbs. 8 oz.			Depth of bar 1·050 in. Breadth „ 1·007 Weight „ 16 tbs. 0½ oz.			Depth of bar 1·034 in. Breadth „ 1·008 Weight „ 16 tbs. 3 oz.		
Weight in tbs.	Deflexion in inches.	Deflexion, load removed.	Weight in tbs.	Deflexion in inches.	Deflexion, load removed.	Weight in tbs.	Deflexion in inches.	Deflexion, load removed.
56	·152	·007	30	·064	+	28	·066	
126	·370	·026	56	·124	·003	56	·137	
182	·570	·060	112	·264	·012	112	·271	·015
238	·797	·092	168	·421	·031	168	·427	·029
294	1·064	·151	224	·588	·054	224	·597	·052
350	1·362	·217	280	·768	·083	230	·780	·080
406	1·710	·318	336	·968	·122	336	·980	·115
448	1·201	·417	392	1·185	·175	392	1·195	·159
472	broke		448	1·434	·253	434	1·370	
			490	1·652		462	1·510	
			504	broke		476	broke	
	ult. defl. =2·168			ult. defl. =1·724			ult. defl. =1·568	

English Irons.

No. 13. Horace St. Paul's, Windmill End iron, No. 2, cold blast, Staffordsh[re].			No. 14. Ley's Works iron, No. 1, hot blast.			No. 15. Lane End iron, No. 2.			No. 16. Carroll iron, No. 2, cold blast.			No. 17. Bierly iron, No. 2, Bradford, Yorkshire.			No. 18. W. S. S. iron, No. 2, Staffordshire.		
Means from 2 experiments.			Means from 3 experiments.			Means from 3 experiments.			Means from 2 experiments.			Means from 3 experiments.			Means from 3 experiments.		
Depth of bar 1·040 in. Breadth „ 1·017 Weight „ 16 lbs.			Depth of bar 1·008 in. Breadth „ 1·012 Weight „ 15 lbs. 4 oz.			Depth of bar 1·005 in. Breadth „ 1·018 Weight „ 15 lbs. 7⅓ oz.			Depth of bar 1·037 in. Breadth „ 1·012 Weight „ 16 lbs. 5 oz.			Depth of bar 1·024 in. Breadth „ 1·036			Depth of bar 1·013 in. Breadth „ 1·007		
Weight in lbs.	Deflexion in inches.	Deflexion, load removed.	Weight in lbs.	Deflexion in inches.	Deflexion, load removed.	Weight in lbs.	Deflexion in inches.	Deflexion, load removed.	Weight in lbs.	Deflexion in inches.	Deflexion, load removed.	Weight in lbs.	Deflexion in inches.	Deflexion, load removed.	Weight in lbs.	Deflexion in inches.	Deflexion, load removed.
56	·112	+	28	·086		28	·070		56	·112	+	28	·061		28	·069	
112	·233	·005	56	·172	·007	56	·140		112	·229	·008	56	·123	+	56	·139	·002
175	·382	·018	112	·369	·031	112	·271	·010	126	·260	·010	112	·258	·011	112	·282	·012
231	·525	·035	168	·597	·065	168	·445	·025	182	·391	·017	168	·384	·021	168	·445	·026
287	·674	·054	224	·850	·107	224	·610	·040	238	·526	·034	224	·533	·039	224	·608	·042
343	·947	·078	280	1·136	·156	280	·780	·057	294	·671	·052	280	·692	·060	280	·789	·062
399	1·117	·108	336	1·447	·227	336	·969	·080	350	·825	·073	336	·853	·086	336	·987	·089
455	1·267	·147	373	1·686	·295	392	1·165	·107	406	·939	·104	392	1·037	·120	383	1·157	·114
511	1·491	·215	404	broke		439	1·339		448	1·120	·125	420	1·136		411	1·264	
532	broke					457	broke		469	broke		439	broke		427	broke	
	ult. defl. =1·519			ult. defl. =1·876			ult. defl. =1·407			ult. defl. =1·187			ult. defl. =1·194			ult. defl. =1·322	

English Irons.

No. 19. Coltham, B. F., iron, No. 1, hot blast, Staffordshire.			No. 20. Corbyn's Hall iron, No. 2, near Dudley, Staffordshire.			No. 21. Wall-Brook iron, No. 3, Dudley, Worcestershire.			No. 22. Oldberry iron, No 3, hot blast, (patent iron) Shropshire.			No. 23. Elsicar iron, No. 1, cold blast.		
Means from 3 experiments.			Means from 3 experiments.			Means from 3 experiments.			Means from 3 experiments.			Means from 2 experiments.		
Depth of bar 1·017 in. Breadth „ 1·009			Depth of bar 1·027 in. Breadth „ 1·022			Depth of bar 1·025 in. Breadth „ 1·033			Depth of bar 1·003 in. Breadth „ 1·001			Depth of bar 1·037 in. Breadth „ 1·027 Weight „ 15 lbs. 10 oz.		
Weight in lbs.	Deflexion in inches.	Deflexion, load removed.	Weight in lbs.	Deflexion in inches.	Deflexion, load removed.	Weight in lbs.	Deflexion in inches.	Deflexion, load removed.	Weight in lbs.	Deflexion in inches.	Deflexion, load removed.	Weight in lbs.	Deflexion in inches.	Deflexion, load removed.
28	·066		28	·069		28	·062		28	·051		56	·140	
56	·133	+	56	·139	+	56	·125	·003	56	·100	+	112	·275	·020
112	·268	·014	112	·288	·015	112	·252	·013	112	·192	+	168	·427	·038
168	·428	·028	168	·466	·036	168	·401	·028	168	·286	·006	224	·591	·054
224	·586	·046	224	·652	·062	224	·556	·049	224	·389	·010	280	·761	·075
280	·756	·068	280	·845	·088	280	·720	·071	280	·492	·014	336	·948	·102
336	·939	·097	336	1·059	·122	336	·902	·103	336	·596	·019	392	1·148	·135
392	1·137	·132	392	1·296	·171	392	1·088	·138	392	·701	·026	448	1·355	·176
439	1·319	·180	448	1·566	·234	451	1·303		448	·804	·031	476	broke	
457	1·393		457	1·613		474	broke		504	·918	·039			
485	broke		464	broke					523	·958	·043			
									546	broke				
	ult. defl. =1·505			ult. defl. =1·643			ult. defl. =1·394			ult. defl. =1·004			ult. defl. =1·450	

Scotch Irons.

No. 1. Carron iron, No. 3, cold blast.			No. 2. Carron iron, No. 3, hot blast.			No. 3. Muirkirk iron, No. 1, cold blast.			No. 4. Muirkirk iron, No. 1, hot blast.			No. 5. Gartsherrie iron, No. 3, hot blast.			No. 6. Dundyvan iron, No. 3, cold blast.			No. 7. Monkland iron, No. 2, hot blast.		
Means from 3 experiments.			Means from 3 experiments.			Means from 2 experiments.			Means from 2 experiments.			Means from 3 experiments.			Means from 3 experiments.			Means from 2 experiments.		
Depth of bar 1·004 in. Breadth ,, 1·005			Depth of bar ·997 in. Breadth ,, 1·006			Depth of bar 1·032 in. Breadth ,, 1·016			Depth of bar 1·020 in. Breadth ,, 1·022			Depth of bar 1·020 in. Breadth ,, 1·025 Weight 15 lbs. 8 oz.			Depth of bar 1·010 in. Breadth ,, 1·018 Weight ,, 15 lbs. 12 oz.			Depth of bar 1·014 in. Breadth ,, ·998 Weight ,, 15 lbs. 4 oz.		
Wt. in lbs.	Deflexion in inches.	Deflexion, load removed.	Wt. in lbs.	Deflexion in inches.	Deflexion, load removed.	Wt. in. lbs.	Deflexion in inches.	Deflexion, load removed.	Wt. in lbs.	Deflexion in inches.	Deflexion, load removed.	Wt. in lbs.	Deflexion in inches.	Deflexion, load removed.	Wt. in lbs.	Deflexion in inches.	Deflexion, load removed.	Wt. in lbs.	Deflexion in inches.	Deflexion, load removed.
28	·069	+	28	·068		28	·071		28	·074		28	·069		28	·067		112	·345	·022
56	·136	·005	56	·129	+	56	·141	+	56	·152	+	56	·145	+	56	·135	+	126	·390	·027
112	·267	·011	112	·249	·006	112	·282	·011	112	·306	·012	112	·292	·018	112	·255	·008	182	·598	·051
168	·418	·024	168	·380	·011	168	·446	·028	168	·476	·025	168	·460	·034	168	·420	·021	238	·827	·083
224	·576	·043	224	·520	·021	224	·624	·051	224	·655	·043	224	·639	·056	224	·579	·037	294	1·075	·118
280	·741	·066	280	·663	·035	280	·814	·081	280	·844	·070	280	·822	·081	280	·746	·059	350	1·358	·174
336	·921	·094	336	·809	·052	336	1·028	·121	336	1·059	·109	336	1·030	·114	336	·932	·089	406	1·693	·273
392	1·114	·129	392	·972	·075	392	1·258	·172	392	1·309	·171	392	1·254	·155	392	1·132	·126	413	broke	
429	1·251		448	1·140	·103	448	1·535	·255	420	1·440		448	1·494	·209	448	1·358	·181			
450	broke		485	1·259		479	broke		444	broke		455	broke		474	broke				
			520	broke																
	ult. defl =1·321			ult. defl. =1·369			ult. defl. =1·682			ult. defl. =1·640			ult. defl. =1·527			ult. defl. =1·456			ult. defl =1·738	

Welsh Irons.

No. 1. Blaina iron, No. 3, cold blast, Monmouthshire.			No. 2. Plaskynaston iron, No. 2, hot blast.			No. 3. Pant iron, No. 2,			No. 4. Beaufort iron, No. 2, hot blast.			No. 5. Beaufort iron, No. 3, hot blast.			No. 6. Maesteg iron, No. uncertain, (marked white) Glamorganshire.		
Means from 3 experiments.			Means from 3 experiments.			Means from 3 experiments.			Means from 3 experiments.			Means from 2 experiments.			Means from 2 experiments.		
Depth of bar 1·039 in. Breadth „ 1·018			Depth of bar 1·009 in. Breadth „ 1·006			Depth of bar 1·034 in. Breadth „ 1·027			Depth of bar 1·037 in. Breadth „ 1·033 Weight „ 16 lbs.			Depth of bar 1·081 in. Breadth „ 1·014 Weight „ 16 lbs. 4½ oz.			Depth of bar 1·041 in. Breadth „ 1·027 Weight „ 16 lbs.		
Weight in lbs.	Deflexion in inches.	Deflexion, load removed.	Weight in lbs.	Deflexion in inches.	Deflexion, load removed.	Weight in lbs.	Deflexion in inches.	Deflexion, load removed.	Weight in lbs.	Deflexion in inches.	Deflexion, load removed.	Weight in lbs.	Deflexion in inches.	Deflexion, load removed.	Weight in lbs.	Deflexion in inches.	Deflexion, load removed.
28	·069		28	·078		28	·063		28	·056		28	·051		28	·062	
56	·135	·006	56	·154	·007	56	·122	·005	56	·116		56	·089	+	56	·131	+
112	·270	·018	112	·320	·021	112	·254	·014	112	·235	·008	126	·231	·004	112	·272	·013
168	·431	·035	168	·511	·040	168	·396	·025	168	·361	·017	182	·346	·012	168	·429	·030
224	·607	·059	224	·706	·064	224	·545	·042	224	·497	·030	238	·467	·024	224	·612	·056
280	·793	·093	280	·915	·092	280	·701	·059	280	·642	·049	294	·592	·037	280	·815	·090
336	1·008	·141	336	1·139	·126	336	·861	·076	336	·797	·073	350	·726	·056	336	1·024	·140
392	1·232	·197	373	1·298	·153	392	1·035	·100	392	·963	·103	406	·872	·078	392	1·262	·191
448	1·535	·290	387	broke		429	1·154	·117	448	1·147	·136	462	1·029	·106	448	1·535	·274
476	broke					448	broke		495	1·258	·183	518	1·200	·142	504	1·880	·415
									513	1·350	·207	574	1·391	·197	504	repeated	
									532	broke		588	1·442	·213		broke	
												599	broke				
	ult. defl. =1·661			ult. defl. =1·354			ult. defl. =1·210			ult. defl. =1·459			ult. defl. =1·479			ult. defl. =1·880	

Welsh Irons.

No. 7. Maesteg iron, No. uncertain, (marked red) Glamorganshire.			No. 8. Pontypool iron, No. 2.			No. 9. Varteg-hill iron, No. 2, hot blast, South Wales.			No. 10. Pentwyn iron, No. 2.			No. 11. Bute iron, No. 1, cold blast.			No. 12. Brimbo iron, No. 2, cold blast.		
Means from 2 experiments.			Means from 2 experiments.			Means from 2 experiments.			Means from 2 experiments.			Means from 3 experiments.			Means from 3 experiments.		
Depth of bar 1·026 in. Breadth „ 1·012 Weight „ 15 lbs. 8 oz.			Depth of bar 1·042 in. Breadth „ 1·020 Weight „ 15 lbs. 12 oz.			Depth of bar 1·017 in. Breadth „ 1·008 Weight „ 15 lbs. 11 oz.			Depth of bar 1·053 in. Breadth „ 1·024 Wt. of 1 of the bars 16 lbs.			Depth of bar 1·027 in. Breadth „ 1·024 Weight „ 15 lbs. 11 oz.			Depth of bar 1·021 in. Breadth „ 1·023 Weight „ 15 lbs. 15 oz.		
Weight in lbs.	Deflexion in inches.	Deflexion, load removed.	Weight in lbs.	Deflexion in inches.	Deflexion, load removed.	Weight in lbs.	Deflexion in inches.	Deflexion, load removed.	Weight in lbs.	Deflexion in inches.	Deflexion, load removed.	Weight in lbs.	Deflexion in inches.	Deflexion, load removed.	Weight in lbs.	Deflexion in inches.	Deflexion, load removed.
28	·070		28	·063		112	·276	·011	28	·056		28	·063		28	·065	
56	·143	+	56	·133	·005	126	·311	·015	56	·116	+	56	·128	+	56	·134	·002
112	·288	·010	112	·291	·020	182	·476	·032	112	·242	·008	112	·262	·010	112	·272	·014
175	·499	·044	168	·457	·038	238	·648	·057	168	·372	·018	168	·416	·022	168	·437	·033
231	·696	·075	224	·641	·065	294	·840	·085	224	·515	·030	224	·578	·042	224	·615	·058
287	·902	·117	280	·838	·097	350	1·045	·123	280	·674	·048	280	·748	·067	280	·805	·088
343	1·170	·186	336	1·045	·142	392	1·220	·169	336	·844	·073	336	·931	·098	336	1·009	·124
399	1·462	·279	392	1·286	·204	420	1·355	·205	392	1·023	·101	392	1·131	·133	392	1·227	·172
455	1·756	·380	448	1·571	·296	434	broke		448	1·221	·139	448	1·357	·185	448	1·477	·237
469	broke		462	1·651					490	1·385		504	1·613	·258	476	1·619	
			486	broke					497	broke		513	1·660		497	broke	
												534	broke				
	ult. defl. =1·840			ult. defl. =1·781			ult. defl. =1·422			ult. defl. =1·410			ult. defl. =1·718			ult. defl. =1·713	

Welsh Irons.

No. 13. Coed-Talon iron, No. 2, cold blast.			No. 14. Coed-Talon iron, No. 2, hot blast.			No. 15. Coed-Talon iron, No. 3, cold blast.			No. 16. Coed-Talon iron, No. 3, hot blast.			No. 17. Ponkey iron, No. 3, cold blast.			No. 18. Frood iron, No. 2, cold blast.		
Means from 3 experiments.			Means from 2 experiments.			Means from 2 experiments.			Means from 2 experiments.			Means from 3 experiments.			Means from 3 experiments.		
Depth of bar 1·048 in. Breadth „ 1·020 Weight „ 16 lbs.			Depth of bar 1·064 in. Breadth „ 1·005 Weight „ 15 lbs. 14 oz.			Depth of bar 1·015 in. Breadth „ 1·011			Depth of bar 1·006 in. Breadth „ 1·003			Depth of bar 1·014 in. Breadth „ 1·022 Weight „ 16 lbs.			Depth of bar 1·011 in. Breadth „ 1·015 Weight „ 15 lbs. 6⅓ oz.		
Weight in lbs.	Deflexion in inches.	Deflexion, load removed.	Weight in lbs.	Deflexion in inches.	Deflexion, load removed.	Weight in lbs.	Deflexion in inches.	Deflexion, load removed.	Weight in lbs.	Deflexion in inches.	Deflexion, load removed.	Weight in lbs.	Deflexion in inches.	Deflexion, load removed.	Weight in lbs.	Deflexion in inches.	Deflexion, load removed.
28	·063		28	·068		28	·063		28	·074	+	28	·056		28	·073	
56	·121	+	56	·130	·005	56	·124	+	56	·146	·004	56	·120	+	56	·144	+
121	·288	·015	126	·327	·027	112	·244	·011	112	·293	·011	112	·241	·010	112	·298	·016
159	·387	·025	182	·505	·054	168	·378	·020	168	·454	·024	168	·373	·020	168	·470	·032
196	·496	·036	238	·699	·087	224	·516	·033	224	·616	·039	224	·515	·031	224	·657	·056
252	·660	·061	294	·910	·122	280	·659	·048	280	·786	·057	280	·657	·048	280	·857	·089
308	·840	·083	350	1·151	·177	336	·806	·066	336	·967	·077	336	·807	·068	336	1·088	·131
364	1·029	·117	406	1·427	·255	392	·966	·089	392	1·156	·101	392	·972	·094	392	1·341	·188
420	1·242	·169	448	1·667	·332	448	1·137	·116	448	1·360	·136	448	1·152	·126	448	1·639	·273
439	1·322	·190	455	1·709		504	1·319	·156	476	1·469		504	1·348	·171	457	1·692	
457	broke		465	broke		546	1·473		504	broke		560	1·572	·233	478	broke	
						560	broke					569	1·613				
												595	broke				
	ult. defl. =1·403			ult. defl. =1·773			ult. defl. =1·521			ult. defl. =1·567			ult. defl. =1·712			ult. defl. =1·798	

Welsh Anthracite Irons.

No. 1. Yniscedwyn Anthracite iron, No. 1, hot blast.			No. 2. Yniscedwyn Anthracite iron, No. 2, hot blast.			No. 3. Yniscedwyn Anthracite iron, No. 3, hot blast.			No. 4. Ystalyfera Anthracite iron, No. 1, hot blast.			No. 5. Ystalyfera Anthracite iron, No. 2, hot blast.			No. 6. Ystalyfera Anthracite iron, No. 3, hot blast.		
Means from 3 experiments.			Means from 3 experiments.			Means from 3 experiments.			Means from 3 experiments.			Means from 3 experiments.			Means from 3 experiments.		
Depth of bar 1·022 in. Breadth ,, 1·013 Weight ,, 15 lbs. 7½ oz.			Depth of bar 1·024 in. Breadth ,, 1·021 Weight ,, 15lbs. 12 oz.			Depth of bar 1·017 in. Breadth ,, 1·010 Weight ,, 15 lbs. 9 oz.			Depth of bar 1·032 in. Breadth ,, 1·050 Weight ,, 15 lbs. 13 oz.			Depth of bar 1·027 in. Breadth ,, 1·038 Weight ,, 15lbs. 10 oz.			Depth of bar 1·020 in. Breadth ,, 1·041 Weight ,, 15lbs. 11 oz.		
Weight in lbs.	Deflexion in inches.	Deflexion, load removed.	Weight in lbs.	Deflexion in inches.	Deflexion, load removed.	Weight in lbs.	Deflexion in inches.	Deflexion, load removed.	Weight in lbs.	Deflexion in inches.	Deflexion, load removed.	Weight in lbs.	Deflexion in inches.	Deflexion, load removed.	Weight in lbs.	Deflexion in inches.	Deflexion, load removed.
28	·072	+	28	·064		28	·063	+	28	·074	+	28	·064	·002	28	·066	+
56	·147	·009	56	·132	+	56	·126	·004	56	·153	·009	56	·134	·006	56	·138	·005
112	·307	·018	112	·263	·013	112	·257	·010	112	·330	·032	112	·281	·025	112	·297	·015
224	·650	·064	224	·558	·039	224	·544	·029	224	·739	·099	224	·623	·068	224	·658	·049
336	1·054	·129	336	·876	·073	336	·854	·063	336	1·258	·210	336	1·021	·147	336	1·065	·107
448	1·545	·250	448	1·241	·139	448	1·197	·111	411	1·694		448	1·500	·263	448	1·550	·216
460	1·601		485	1·366		504	1·385	·146	448	1·944	·458	476	1·641		495	broke	
477	broke		504	1·441		538	broke		457	2·009		495	broke				
			518	broke					483	broke							
	ult. defl. =1·693			ult. defl. =1·497			ult. defl. =1·499			ult. defl. =2·181			ult. defl. =1·742			ult. defl. =1·792	

94. *Mean results of experiments made, by the Author, on the transverse strength and elasticity of uniform bars of cast iron, of different forms of section. All the bars, except the last, No.* 16, *being cast* 5 *feet long, and laid on supports* 4 *feet* 6 *inches asunder, and having the weights suspended from the middle. The results are abridged from the experiments in the Author's Report on the Strength, &c., of Hot and Cold Blast Iron* (Brit. Association, vol. vi.), *and from other experiments recently made.*

RECTANGULAR BARS OF ENGLISH IRON.									RECTANGULAR BARS.			RECTANGULAR BARS OF SCOTCH IRON.		
No. 1. Buffery iron, No. 1, hot blast, near Birmingham.			No. 2. Buffery iron, No. 1, cold blast, near Birmingham.			No. 3. Low Moor iron, No. 3, cold blast, Yorkshire.			No. 4. Mixture of iron used in beams of Liverpool and Leeds Junction Railway, at Salford.			No. 5. Carron iron, No. 2, hot blast, made with coke.		
Means from 3 experiments.			Means from 3 experiments.			Means from 2 experiments, *Phil. Trans.*, Part II., 1840.			From 1 experiment.			Means from 3 experiments.		
Depth of bar 1·002 in. Breadth „ 1·017 Weight „ 15 lbs. 8 oz.			Depth of bar 1·009 in. Breadth „ ·980 Weight „ 15 lbs. 8 oz.			Depth of bar 1·003 in. Breadth „ 1·000			Depth of bar 1·030 in. Breadth „ 1·025 Weight „ 15 lbs. 13 oz.			Depth of bar 1·004 in. Breadth „ 1·004 Weight „ 15 lbs. 5 oz.		
Weight in lbs.	Deflexion in inches.	Deflexion, load removed.	Weight in lbs.	Deflexion in inches.	Deflexion, load removed.	Weight in lbs.	Deflexion in inches.	Deflexion, load removed.	Weight in lbs.	Deflexion in inches.	Deflexion, load removed.	Weight in lbs.	Deflexion in inches.	Deflexion, load removed.
112	·316	·014	112	·278	·008	28	·065	·001 ?	28	·055	+	16	·037	visible
224	·687	·046	224	·603	·044	56	·143	·005	56	·130	increased	23	·052	„
336	1·150	·116	336	·984	·096	112	·314	·025	112	·290	·025	30	·069	·001 ?
392	1·420	·183	392	1·207	·142	224	·701		224	·63	·07	56	·131	·002
411	1·520	·211	448	1·45	·199	336	1·186	·17	336	1·55	·14	112	·269	·008
446	broke		476	broke		392	1·481	·26	392	1·83	·23	224	·582	·036
						448	1·826	·38				336	·933	·086
						467	broke					448	1·348	·174
												469	broke	
	ult. defl. =1·647			ult. defl. =1·54			ult. defl. =1·943						ult. defl. =1·430	

Rectangular Bars of Scotch Iron.

No. 6. Carron iron, No. 2, cold blast, made with coke.			No. 7. Devon iron, No. 3, hot blast, made with coal.			No. 8. Devon iron, No. 3, cold blast, made with coke.			No. 9. Carron iron, No. 2, bars cast $1\frac{1}{4}$ inch square, and reduced in middle to 1 inch, nearly.			No. 10. Carron iron, No. 2.			No. 11. Carron iron, No. 2.		
Means from 3 experiments.			Means from 2 experiments.			Means from 2 experiments.			Means from 4 expts., 2 on hot & 2 on cold blast iron.			Means from 2 expts., 1 on hot & 1 on cold blast iron.			Means from 3 expts., 2 on hot & 1 on cold blast iron.		
Depth of bar 1·029 in. Breadth „ 1·014 Weight „ 15 lbs. 12 oz.			Depth of bar 1·005 in. Breadth „ 1·005			Depth of bar 1·00 in. Breadth „ 1·00			Depth of bar in middle— 1·017 in. Breadth „ „ 1·018			Depth of bar 3·000 in. Breadth „ 1·025 Weight $46\frac{1}{2}$ lbs. nearly.			Depth of bar 4·977 in. Breadth „ 1·023 Weight „ 77 lbs. 11 oz.		
Weight in lbs.	Deflexion in inches.	Deflexion, load removed.	Weight in lbs.	Deflexion in inches.	Deflexion, load removed.	Weight in lbs.	Deflexion in inches.	Deflexion, load removed.	Weight in lbs.	Deflexion not taken.	Deflexion, load removed.	Weight in lbs.	Deflexion in inches.	Deflexion, load removed.	Weight in lbs.	Deflexion in inches.	Deflexion, load removed.
16	·033	visible?	112	·192		112	·192		112		·006	1082	·088	·002	4936	·107	·01 nearly
30	·062	„	168	·295	·002	168	·292		140		·010	1343	·109	·004	5867	·130	·01
56	·117	·003	224	·405	·008	224	·395		168		·022	1605	·134	·006	6798	·153	·015
112	·230	·006	280	·505		280	·495		196		·030	1866	·160	·008	7730	·182	·02
168	·358	·012	336	·615	·020	336	·590	·004	224		·043	2126	·188	·011	8662	·208	·02
224	·490	·022	392	·730		392	·690		336		·118	2388	·216	·014	9593	·227	·03
280	·621	·034	448	·845	·045	420	·740		392		·163	2649	·246	·018	10298	broke	
336	·762	·053	504	·980		448	broke		448		·220	2910	·276	·024			
392	·913	·075	539	1·070					476	broke		3172	·309	·030			
448	1·073	·105	546	broke								3433	·342	·037			
511	broke											3694	·378	·048			
												3890	broke				
	ult. defl. =1·278			ult. defl. =1·08			ult. defl. =·790						ult. defl. =·406			ult. defl. =·265	

Bars of Scotch Iron, Carron, No. 2.

No. 12. Bar, section an isosceles triangle, base 1·43 inches, and each side 2·00 inches, broken with the vertex downwards. ▽			No. 13. Bar same as last, and filed to the exact size, but having $\frac{1}{10}$th of the depth taken off the vertex. It was broken with the vertex downwards. ▽			No. 14. Uniform bar, whose form of section is below, broken with the vertical rib downwards. ┬			No. 15. Bar same as last, broken with the vertical rib upwards. ┴			No. 16. Larger bar, from same model as those in page 381, broken with the vertical rib upwards. ┴		
Means from 3 experiments, 2 on hot and 1 on cold blast iron.			Means from 2 experiments, one on hot and the other on cold blast iron.			Means from 2 experiments, one on hot and the other on cold blast iron.			Means from 2 experiments, one on hot and the other on cold blast iron.			Means from 2 experiments, distance between supports 6 feet 6 inches. A fluid iron, but its name unknown.		
Weight in lbs.	Deflexion in inches.	Deflexion, load removed.	Weight in lbs.	Deflexion in inches.	Deflexion, load removed.	Weight in lbs.	Deflexion in inches.	Deflexion, load removed.	Weight in lbs.	Deflexion in inches.	Deflexion, load removed.	Weight in lbs.	Deflexion in inches.	Deflexion, load removed.
112	·07		112	·072		112	·195		112	·205		56	·112	
168	·106		168	·110	·002	168	·305	·013	224	·40	·02	112	·238	·014
224	·145	·005	224	·150	·003	196	·365		280	·50	·025	224	·498	·033
336	·227	·010	336	·235	·006	224	·425	·025	336	·60		336	·776	·072
448	·317	·020	448	·330	·014	252	·495	·040	392	·715	·065	448	1·050	·114
560	·420	·040	560	·430	·030	273	broke		448	·83		560	1·401	·181
672	·530		658	·520					560	1·105	·165	672	1·720	·261
766	broke		702	broke					672	1·43		784	2·122	·397
									784	1·84		896	2·542	·592
									896	2·425		1008	3·130	·872
									994	3·050		1120	3·892	1·275
									1015	broke		1176	4·290	
												1232	broke	
	ult. defl. =·62			ult. defl. =·58			ult. defl. =·54			ult. defl. =3·19			ult. defl. =4·757	
									Fracture attended by the separation of a wedge of the form			Fracture caused by tension without the separation of a wedge.		

REMARKS ON THE EXPERIMENTS IN THE LAST ARTICLE.

95. The objects sought for in every experiment were to obtain—the deflexion of the beam with given weights, generally increasing by equal increments,—and the set, or defect of elasticity, as exhibited by the deviation of the beam from its original form, after the weight was taken off; for reasons before mentioned (art. 86). As the beams were generally broken, the deflexion at the time of fracture could not often be obtained by direct admeasurement; it was therefore usually calculated, for each separate experiment, from the breaking weight, and the last weight, with its observed deflexion, previous to fracture.

Many of the experiments were on the Carron Iron, No. 2, both hot and cold blast; and as these two denominations of iron differ in transverse strength only as 99 to 100, from the results of a great number of experiments, we may consider them as of the same strength. Some of the experiments were made to test admitted conclusions with respect to the strength of materials, and their objects will now be described.

The experiments Nos. 5 and 6 were on bars somewhat greater than 1 inch square; and the bars Nos. 10 and 11 were nearly of the same breadth as those, but had their depths 3 and 5 inches respectively. Hence, supposing each bar to be 1 inch broad, and the depths 1, 3, 5 inches respectively, the strengths should be as 1, 9, 25, the square of the depths. The mean strength from the bars Nos. 5 and 6 was 490lbs., but as these were larger than 1 inch square, I will state the reduced results from a number of experiments. The strength per square inch from five cold blast bars of this iron was 467 lbs., and from five hot blast bars it was 459 lbs.; mean from the whole 463 lbs. Multiplying this mean by 9 and 25, gives for the strength of the bars, 3 and 5 inches deep, 4167 and 11,575 lbs. respectively. The 3 and 5-inch bars, in Nos. 10 and 11, are rather more than 1 inch broad, and their mean strengths are 3890 and 10,298 lbs. And, if the breadths were reduced to 1 inch, the strengths would be 3795, 10,067 lbs. respectively. But, as these numbers are less than 4167, 11,575, as above, it appears that the strength of rectangular bars, of the same length and breadth, increases in a ratio somewhat lower than as the square of the depth.

It having been often asserted that, if the external part or crust of a cast iron bar be taken away, the strength of the internal part will be much less than that of a bar of the same dimensions, retaining its outer crust; the result of the experiments in No. 9, compared with those of Nos. 5 and 6, will show that the falling off in strength, if any, is not great.

It was asserted by Emerson, in his 'Mechanics,' 4to, page 114, that if a beam be made in the form of a triangular prism, and $\frac{1}{9}$th of the height be taken from the vertex, parallel to the base, the remaining part will be stronger than the whole beam: this result was obtained on a supposition that materials are incompressible. Tredgold, in art. 118 of this volume, computes the same, on the supposition of equal extensions and compressions from equal forces, and finds that, if $\frac{1}{10}$th of the depth of such a beam be taken away from the vertex, the strength will be about the greatest. The experiments Nos. 12 and 13 were intended to show how far this was true; and it appears that the frustrum, with $\frac{1}{10}$th of the depth taken away, instead of being stronger than the whole triangle, was weaker than it in the proportion of 702 to 766. The object of the experiments Nos. 14 and 15 was principally to show that beams of cast iron, of the same dimensions, might be made to bear, when turned one way upwards, several times the weight which they would bear when turned the opposite way up. In this case the strengths were as 1015 to 273, or as 4 to 1 nearly. This was first shown in the Author's Paper on the 'Strength and best Form of Iron Beams' (Manchester Memoirs, vol. v.), and of which an abstract will be given in this volume. The experiments No. 16 have in part the same object as those in art. 86, before described.

96. In the following Table, the result from each bar is reduced to exactly 1 inch square; and the transverse strength, which may be taken as a criterion of the value of each iron, is obtained from a mean between the reduced results of the original experiments upon it;—first on bars 4 feet 6 inches between the supports, and next on those of half the length, or 2 feet 3 inches between the supports. All the other results are deduced from the 4 ft. 6 inch bars. In all cases the weights were laid on the middle of the bar.

Abstract of Results obtained from the whole of the Experiments, both of Mr. Fairbairn and the Author, on the Transverse Strength, and other properties, of cast iron bars, from the principal Iron Works in the United Kingdom.

Number of iron in the scale of strength.	Names of irons.	Number of experiments on each.	Specific gravity.	Modulus of elasticity in lbs. per square inch, or stiffness.	Breaking weight in lbs. of bars 4 ft. 6 in. between supports.	Breaking weight in lbs. of bars 2 ft. 3 in. reduced to 4 ft. 6 in. between supports.	Mean breaking weight in lbs. (S.)	Ultimate deflexion of 4 ft. 6 in. bars, in parts of an inch.	Power of the 4 ft. 6 in. bars to resist impact.	Colour.	Quality.
1	Ponkey, No. 3, cold blast	4	7·122	17211000	567	590	578	1·747	992	Whitish gray	hard
2	Devon, No. 3, hot blast*	2	7·251	22473650	537	—	537	1·09	589	White	hard
3	Oldberry, No. 3, hot blast	5	7·300	22733400	543	517	530	1·005	549	White	hard
4	Carron, No. 3, hot blast	2	7·056	17873100	520	534	527	1·365	711	Whitish gray	hard
5	Coed-Talon, No. 3, hot blast	4	6·970	14707900	496	530	518	1·577	782	Dullish gray	hard
6	Beaufort, No. 3, hot blast	5	7·069	16802000	505	529	517	1·599	807	Dullish gray	hard
7	Butterley	4	7·038	15379500	489	515	502	1·815	889	Dark gray	soft
8	Bute, No. 1, cold blast	4	7·066	15163000	495	487	491	1·764	872	Bluish gray	soft
9	Wind Mill End, No. 2, cold blast	4	7·071	16490000	483	495	489	1·581	765	Dark gray	hard
10	Old Park, No. 2, cold blast	5	7·049	14607000	441	529	485	1·621	718	Gray	soft
11	Carron, No. 2, cold blast*	3	7·066	17270500	476	—	476	1·313	630	Dull gray	rather hard
12	Beaufort, No. 2, hot blast	4	7·108	16301000	478	470	474	1·512	729	Dull gray	hard
13	Low Moor, No. 2, cold blast	4	7·055	14509500	462	483	472	1·852	855	Dark gray	soft
14	Low Moor, No. 3, cold blast*	2	7·052	13918740	467	—	467	1·944	908	Dark gray	rather harder
15	Buffery, No. 1, cold blast*	3	7·079	15381200	463	—	463	1·550	721	Gray	rather hard
16	Carron, No. 2, hot blast*	3	7·046	16085000	463	—	463	1·337	619	Grayish blue	rather hard
17	Brimbo, No. 2, cold blast	5	7·017	14911666	466	453	459	1·748	815	Light gray	rather hard
18	Apedale, No. 2, hot blast	3	7·017	14852000	457	455	456	1·730	791	Light gray	stiff
19	Oldberry, No. 2, cold blast	4	7·059	14307500	453	457	455	1·811	822	Dark gray	rather soft
20	Pentwyn, No. 2	4	7·038	15193000	438	473	455	1·484	650	Bluish gray	hard
21	Maesteg, No. 2	5	7·038	13959500	453	455	454	1·957	886	Dark gray	rather soft
22	Muirkirk, No. 1, cold blast	4	7·113	14003550	443	465	454	1·734	770	Bright gray	fluid
23	Adelphi, No. 2, cold blast	5	7·080	13815500	441	457	449	1·759	777	Light gray	soft
24	Blaina, No. 3, cold blast	5	7·159	14281466	433	464	448	1·726	747	Bright gray	hard
25	Devon, No. 3, cold blast*	2	7·295	22907700	448	—	448	·790	354	Light gray	hard
26	Gartsherrie, No. 3, hot blast	5	7·017	13894000	427	467	447	1·557	998	Light gray	soft

27	Frood, No. 2, cold blast	5	7·031	14112666	460	434	447		1·825	841		Light gray	open
28	Lane End, No. 2	3	7·028	15787666	444	—	444		1·414	629		Dark gray	soft
29	Carron, No. 3, cold blast*	5	7·094	16246966	444	448	446		1·336	594		Gray	soft
30	Dundyvan, No. 3, cold blast	4	7·087	16534000	456	430	443		1·469	674		Dull gray	rather soft
31	Maesteg (marked red)	5	7·038	13971500	440	444	442		1·887	830		Bluish gray	fluid
32	Corbyn's Hall, No. 2	5	7·007	13845866	430	454	442		1·687	727		Gray	soft
33	Pontypool, No. 2	5	7·080	13136500	439	441	440		1·857	816		Dull blue	rather soft
34	Wallbrook, No. 3	5	6·979	15394766	432.	449	440		1·443	625		Light gray	rather hard
35	Milton, No. 3, hot blast	4	7·051	15852500	427	449	438		1·368	585		Gray	rather hard
36	Buffery, No. 1, hot blast*	3	6·998	13730500	436	—	436		1·640	721		Dull gray	soft
37	Level, No. 1, hot blast	5	7·080	15452500	461	403	432		1·516	699		Light gray	soft
38	Pant, No. 2	5	6·975	15280900	408	455	431		1·251	511		Light gray	rather hard
39	Level, No. 2, hot blast	6	7·031	15241000	419	439	429		1·358	570		Dull gray	soft
40	W. S. S., No. 2	5	7·041	14953333	413	446	429		1·339	554		Light gray	soft
41	Eagle Foundry, No. 2, hot blast	4	7·038	14211000	408	446	427		1·512	618		Bluish gray	soft
42	Elsicar, No. 2, cold blast	4	6·928	12586500	446	408	427		2·224	992		Gray	soft
43	Varteg, No. 2, hot blast	4	7·007	15012000	422	430	426		1·450	621		Gray	hard
44	Coltham, No. 1, hot blast	5	7·128	15510066	464	385	424		1·532	716		Whitish gray	rather soft
45	Carroll, No. 2, cold blast	4	7·069	17036000	430	408	419		1·231	530		Gray	hard
46	Muirkirk, No. 1, hot blast	4	6·953	13294400	418	420	419		1·570	656		Bluish gray	soft
47	Bierley, No. 2	5	7·185	16156133	404	432	418		1·222	494		Dark gray	soft
48	Coed-Talon, No. 2, hot blast	4	6·969	14322500	409	424	416		1·882	772		Bright gray	soft
49	Coed-Talon, No. 2, cold blast	5	6·955	14304000	408	418	413		1·470	600		Gray	rather soft
50	Monkland, No. 2, hot blast	3	6·916	12259500	402	404	403		1·762	709		Bluish gray	soft
51	Ley's Works, No. 1, hot blast	3	6·957	11539333	392	—	392		1·890	742		Bluish gray	soft
52	Milton, No. 1, hot blast	4	6·976	11974500	353	386	369		1·525	538		Gray	soft and fluid
53	Plaskynaston, No. 2, hot blast	5	6·916	13341633	378	337	357		1·366	517		Light gray	rather soft
	ANTHRACITE IRONS.												
54	Yniscedwyn Anthracite, No. 1, hot blast	6	7·078	13741400	453	464	458		1·730	785		Grayish blue	soft
55	" 2, "	5	7·095	15334000	485	532	508		1·529	709		Grayish blue	harder
56	" 3, "	5	7·168	16194327	515	525	520		1·525	785		Whitish gray	rather harder
57	Ystalyfera Anthracite, First sample, No. 1, hot blast	6	6·992	11555635	435	423	429	411 mean	2·252	973	771 mean	Bluish gray	
	Second do. " "	4	7·098	14044420	392·3		392·3		1·445	569			
58	First do. No. 2, "	6	7·053	13973270	453	454	454	467	1·788	810	769	Dark gray	rather soft
	Second do. " "	4	7·258	15686750	480·7		480·7		1·505	728			
59	First do. No. 3, "	6	7·133	13436806	457	475	466	484	1·825	837	751	Whitish gray	
	Second do. " "	4	7·352	18391425	502		502		1·324	665			

* The irons marked with an asterisk are from the Author's Experiments.

97. Since the experiments above were given to the public, some others, upon bars of the same dimensions, and having their results reduced in the same manner as these, have been published by Mr. David Mushet. Other experiments on the Ystalyfera iron have been given to the public by Mr. Evans: those above are results obtained from experiments upon two samples of each kind, sent to Mr. Fairbairn from the proprietors.

Mr. Fairbairn has likewise recently sent to the Institution of Civil Engineers the results of experiments made for him by the Author, upon bars of the same size as those in the preceding pages, and on four other kinds of cast iron, viz.: iron obtained from Turkish ores; iron from the island of Elba; and two kinds of Ulverston (English) iron.

98. *Explanation and Uses of the preceding Table.*

1st. *Explanation.*—The column representing the number of experiments refers to those from which the strength of the beams was obtained.

The specific gravity was obtained, generally, from a mean of about half a dozen experiments, on small specimens, weighed in and out of water.

The modulus of elasticity was usually obtained from the deflexion caused by 112 ℔s. on the 4 feet 6 inch bars, calculated from the value of m in the formula, $m=\frac{w\,l^3}{4\,b\,d^3\,a}$, (Part I. art. 256). The numbers representing the power to resist impact were obtained from the product of the breaking weight of the bars, by their ultimate deflexion; as it appeared from the experiments in the Author's Paper 'On Impact upon Beams,' (British Association of Science, fifth Report,) that the conclusions of Tredgold (art. 304), with respect to a modulus of resilience, applicable so long as the elasticity was uninjured, might be extended to the breaking point in cast iron.

2nd. *Uses of the Table.*—These are numerous, but two only of the most common will be mentioned. If b and d be the breadth and depth of a rectangular beam in inches, l the distance between the supports in feet, w the breaking weight in lbs., w' any other weight, d' its deflexion, and m the modulus of elasticity in lbs., for a square inch: putting 4·5 for the distance 4 feet 6 inches, above, we have

$$w = \frac{4{\cdot}5 \times b\, d^2\, s}{l}$$ { The value of s being taken from the Table above.

$$w' = \frac{m\, b\, d^3\, d'}{432\, l^3}$$ { (Part I. art. 256,) the value of the modulus m being obtained from the Table.

DEFECT OF ELASTICITY.

99. In all the preceding experiments on rectangular bars, the defect of elasticity, measured by the deflexion remaining in the bar after the load had been removed, was observed, for reasons previously given (arts. 86, 92); and to show the law which regulates this defect, its value, with equal additions of weight, will be collected from the mean results upon each iron, and placed under the corresponding weights in the following Table.

100. *Defect of elasticity, or set, as obtained from the mean deflexion of bars cast from models* 1 *inch square, laid on supports* 4·5 *ft. asunder; using only those irons upon which experiments had been made, as to the set, upon all the weights set down.*

FIRST SERIES.	56	112	168	224	280	336	392	448
No. 1 Irons.								
Elsicar, cold blast		·020	·038	·054	·075	·102	·135	·176
Muirkirk, cold blast		·011	·028	·051	·081	·121	·172	·255
Bute, cold blast		·010	·022	·042	·067	·098	·133	·185
No. 2 Irons.								
Corbyn's Hall		·015	·036	·062	·088	·122	·171	·234
Beaufort, hot blast		·008	·017	·030	·049	·073	·103	·136
Pentwyn		·008	·018	·030	·048	·073	·101	·139
Frood, cold blast		·016	·032	·056	·089	·131	·188	·273
No. 3 Irons.								
Carron, hot blast		·006	·011	·021	·035	·052	·075	·103
Gartsherrie, hot blast . . .		·018	·034	·056	·081	·114	·155	·209
Dundyvan, cold blast		·008	·021	·037	·059	·089	·126	·181
Coed-Talon, do.		·011	·020	·033	·048	·066	·089	·116
Ponkey, do.		·010	·020	·031	·048	·068	·094	·126
Maesteg, number unknown . .		·013	·030	·056	·090	·140	·191	·274
No. 2 Irons.								
Oldberry, cold blast	·003	·012	·031	·054	·083	·122	·175	·253
Pontypool	·005	·020	·038	·065	·097	·142	·204	·296
Brimbo	·002	·014	·033	·058	·088	·124	·172	·237
Carron, cold blast	·003	·006	·012	·022	·034	·053	·075	·105
No. 3 Irons.								
Blaina, cold blast	·006	·018	·035	·059	·093	·141	·197	·290
Coed-Talon, hot blast . . .	·004	·011	·024	·039	·057	·077	·101	·136
Means from sets from nineteen kinds of iron		·0124	·026	·045	·069	·100	·140	·196
Sets computed from the formula $x=\frac{W^2}{342}$, where x is the set, and w the weight in ½ cwts.		·0117	·026	·047	·073	·105	·143	·187
SECOND SERIES.								
No. 2 Irons.								
Adelphi, cold blast	·002	·014	·034	·060	·093	·138	·201	
Eagle Foundry, hot blast . .	·003	·013	·030	·051	·078	·113	·159	
Level, hot blast	·002	·011	·022	·038	·061	·088	·121	
Pant	·005	·014	·025	·042	·059	·076	·100	
No. 3 Irons.								
Wallbrook	·003	·013	·028	·049	·071	·103	·138	
Carron, cold blast	·005	·011	·024	·043	·066	·094	·129	
Means from the six kinds of iron	·0037	·0127	·027	·047	·071	·102	·141	
Means from the last six in former part of Table . . .	·0038	·0128	·029	·051	·080	·115	·157	
Means from the twelve kinds of iron	·0037	·0127	·028	·049	·076	·109	·149	
Sets computed, as before, from formula $x=\frac{W^2}{328}$, x and w being as above	·0030	·0122	·027	·049	·073	·110	·149	

101. Comparing the mean sets, or defects of elasticity, in each series of the preceding Table, with the computed ones, it appears that the defects vary nearly as the square of the weights; the set being the abscissa and the weight the ordinate of a parabola.

Hence there is no force, however small, that will not injure the elasticity of cast iron.

102. When bars of a ⊥ form of section are bent, so as to make the flexure to depend upon the extension or compression of the vertical rib (as in arts. 86, and 94, experiments 14 to 16), the set is nearly as the square of the extension or compression; these being measured by the deflexions.

103. In all the preceding experiments, the weight laid upon the beam acted in a vertical direction; and the weight of the beam, independently of the other weight, had a small tendency to deflect the beam; the deflexions given in the Table being measured as commencing from that position which the beam had taken in consequence of its own weight. This, therefore, introduced an error which, though very small, on account of the great strength of cast iron compared with its weight, ought, if possible, to be avoided; especially where the object was,—not only to prove that defects of elasticity were produced by weights which were not hitherto supposed capable of injuring the elasticity,—but also to seek for the law which regulated these defects. Other objections to these results might be urged, as for instance: when a beam is laid upon two

supports, and bent by a weight in the middle or elsewhere; the friction between the ends of the beam and the supports will have a slight influence upon the deflexion, a matter which has been submitted to calculation by Professor Moseley[10] in his able work on engineering. To meet the objections above, I had an apparatus constructed with four friction wheels, two to support each end of the beam; one wheel acting horizontally and the other vertically. The horizontal wheels were intended to destroy the friction arising from the weight of the beam, and the vertical ones that from the weight applied; this weight, in its descent, being made to act horizontally upon the beam, by means of a cord passing over a pulley. The results obtained in this way confirm the truth of the former ones; and by being freed from small errors, are much more consistent among themselves than they would otherwise have been.

104. A bar of the ⊥ form of section, bent so as to compress the vertical rib, with weights varying from 112 to 1344 lbs., gave, from a mean of two experiments very carefully made, the set, as the 1·88 power of the deflexion, measuring the compression of the rib. In these experiments, each weight was allowed to remain on the beam five minutes, and the set was taken twice, at intervals of one and five minutes after unloading; it having been found that

10 'Mechanical Principles of Engineering and Architecture,' art. 389.

a greater length of time produced but little change in the quantity of the set.

105. Experiments made to extend the vertical rib, the bar, during flexure, being turned the opposite side upwards, gave the set as a power of the deflexion, or extension, somewhat higher than as above.

106. Supposing the set to arise wholly from the extension or compression of the rib, which is very probable, it will therefore be nearly as the square of the extension or compression, as above observed. If, therefore, x represent the quantity of extension or compression, which a body has sustained, and $a x$ the force producing that extension or compression, on the supposition that the body was perfectly elastic; then, the real force f, necessary to produce the extension or compression x, will be smaller, than on the supposition of perfect elasticity, by a quantity $b x^2$; and we shall have $f = a x - b x^2$.

107. The law of defective elasticity, as here given, and its application to other materials, as stone, timber, &c., was discovered by the author in July, 1843, and laid before the British Association of Science, at its meeting in Cork.

OF THE SECTION OF GREATEST STRENGTH IN CAST IRON BEAMS.

108. The very extensive and increasing use made of cast iron beams renders it exceedingly desirable that they should be cast in the form best suited for insuring strength; and that, if possible, formulæ

should be obtained by which the strength can be estimated. Without these the engineer and founder must be in constant uncertainty; and either endanger the stability of erections, costing many thousands of pounds, and perhaps supporting hundreds of human beings, or incur the risk of employing an unnecessary quantity of metal, which, besides its expense, does injury by its own weight.

109. The earliest use of this most valuable material for beams has been but of recent date: so far as I can learn it was first used by Boulton and Watt, who in 1800 employed beams, whose section was of the form ⊥, in building the cotton mill for Messrs. Philips and Lee in Salford. These, the earliest cast iron beams, differed from the ⊥ formed beams of the present day, in having the lower portion of the vertical part thicker than the modern ones. Both have had the same object in their construction, that of supporting arches of brick-work for the floors of fire-proof buildings; and as they were well suited for that purpose, and of a convenient form for casting, besides being very strong, particularly the modern ones, comparatively with rectangular beams of that metal, their use has been very general, and they are still employed; though they have now been supplanted in most of the large erections of Manchester and its neighbourhood, and many other parts of the kingdom, by another form derived from experiments of which I gave an account in the fifth volume of the 'Memoirs of the

Literary and Philosophical Society of Manchester' (second series), published in 1831. I propose giving here extracts from the leading results and reasonings in that Paper.

110. In the application of a material like cast iron to purposes to which it had not been before applied, it could not be expected that the form best suited for resistance to strain, any more than the quantity necessary to support that strain, could be at once attained. The ⊥ form of cast iron beam mentioned above, was, however, by no means a bad one; it had undergone modifications and improvements by different parties, and had had various experiments made upon it, some of which, made by my friend Mr. Fairbairn, on a large scale, I gave in the Paper mentioned above. This form, however, Mr. Tredgold saw, was not the best, and gave, in his article on the 'Strongest Form of Section' (Part I. Section IV.), a representation of what he considered the best (Plate I. fig. 9), a beam with two equal ribs or flanges, one at the top and the other at the bottom. Mr. Tredgold proposed this form, assuming that, whilst the elasticity of a body is perfect, it resists the same degree of extension or compression with equal forces; and therefore he concluded that a beam, to bear the most, should have equal ribs at top and bottom, as it ought not to be strained so as to injure its elasticity.[11]

[11] Mr. Tredgold did not suspect that the elasticity was injured by forces however small, see art. 86, &c.

111. Having myself given a Paper on the 'Transverse Strength of Materials,' in the fourth volume of the 'Memoirs of the Literary and Philosophical Society of Manchester,' published in 1824, containing the mathematical developement of some principles to which I attached importance, besides some experiments to ascertain the position of the neutral line in bent pieces of timber, I felt persuaded that the form proposed by Mr. Tredgold was not the best to resist fracture in cast iron. It was evident that that metal resisted fracture by compression with much greater force than it did by tension, though the ratio was then unknown; and I was convinced that the transverse strength of a bar depended in some manner upon both of these forces; the situation of the neutral line being changed before fracture in consequence of their inequality.

112. To obtain further information on this subject, I adopted, about the year 1828, a mode of analyzing, separately, the forces of extension and compression, in a bent body; results of which have been given in arts. 86 and 94, from experiments, since made for other purposes. I had bars cast from a model, 5 feet long, whose section was of the form

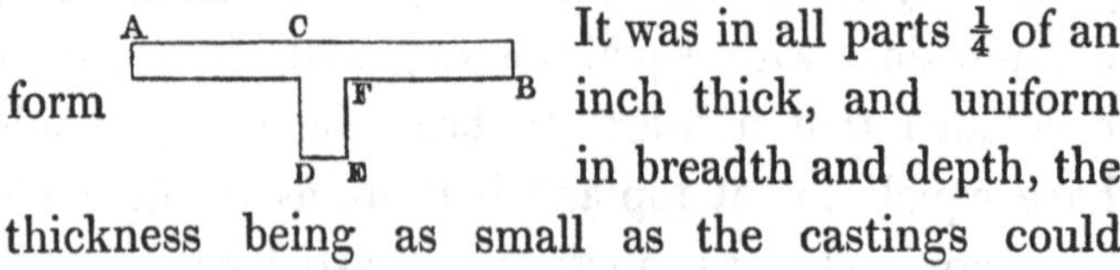

It was in all parts $\frac{1}{4}$ of an inch thick, and uniform in breadth and depth, the thickness being as small as the castings could be run to make them uniform and sound; the breadth AB of the flange was 4 inches, and the

depth F E of the rib running along its middle was 1·1 inch.

113. When the castings so formed had their ends placed horizontally upon supports, and weights were suspended from the middle, the flexure would depend almost entirely upon the contraction or extension of the rib F E. When the rib was upwards, the deflexion would arise from the contraction of that rib, and when downwards, from its extension.

114. To ascertain the resistance to fracture in these two cases, I took two castings, apparently precisely alike, and placing the ends of each of them upon two props 4 feet 3 inches asunder, broke them by weights in the middle, one with the rib upwards, and the other with it downwards, as in the figure.

That with the rib downwards bore $2\frac{1}{4}$ cwt., and broke with $2\frac{1}{2}$ cwt. The other casting bore $8\frac{3}{4}$ cwt., and broke with 9 cwt. Deflexion of the latter in middle with 4 cwt. $=$ ·6 inch, with $8\frac{1}{4}$ cwt. $=$ 1·8 inch.

115. The strength of the castings was, therefore, nearly as $2\frac{1}{2}$ to 9, or as 10 to 36, accordingly as they were broken one or the other way upwards.

116. When the second broke a piece flew out, whole, from the compressed side of the casting, of the following form A, D, B, where A B $=$ 4 inches, and C D $=$ ·98 inch; the point D at the bottom being in or near to the neutral line of the bar, a side view of which is represented in the figure.

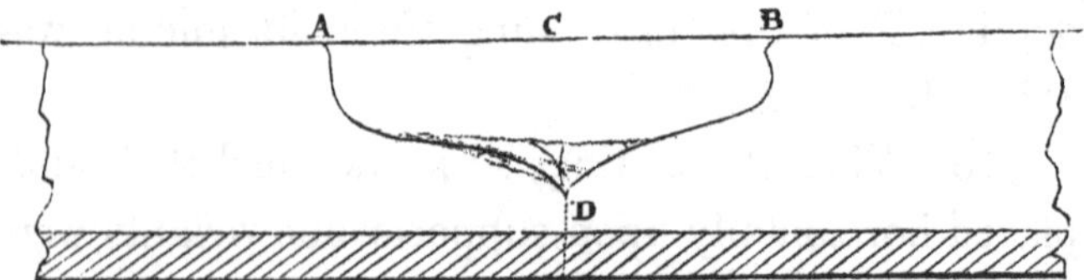

The side A B of the wedge-like piece broken out, was, as will be seen, in the direction of the length of the casting, and the weights were laid on at C. Hence, as the depth of the casting was found to be 1·35 inch, $CD = \frac{\cdot 98}{1\cdot 35} = \frac{10}{14}$ of the depth, nearly. In the experiment on the second bar, (art. 86,) made since that time, $CD = \frac{1\cdot 05}{1\cdot 56} = \frac{10}{15}$ nearly.

117. These experiments are interesting; they show the effect of the position of the casting on the strength; give the situation of the neutral line; and may, from the peculiar form of the wedge, which, as represented here, is more perfect than usual, throw some additional light on the nature of the strain.

118. Those who with Tredgold (Part I. art. 37, &c.) suppose the strength to be bounded by the elasticity, and that the same force would destroy the elastic power, whether it was applied to extend or compress the body, must have conceived these castings, and indeed those of every other form, to be equally strong, whichsoever way upwards they were turned;—a conclusion which we see would lead to very erroneous results, if applied to measure the ultimate strength of cast iron.

Other experiments were made at that time upon bars of the same form as the preceding, to ascertain the deflexions with given forces when the rib C D was subjected alternately to tension and compression; and it was shown, as might be expected, that the extensions and compressions, measured by the deflexions, were nearly equal from the same forces, though the extension was usually somewhat greater than the compression, the difference increasing with the weight, through the whole range to fracture. This always took place by the rib being torn asunder, the compression necessary to produce fracture being several times as great as the extension required to do it, as may be inferred from art. 33.

119. The object of the preceding experiments being to prepare, in some degree, the way to an inquiry into the best forms of beams of cast iron; we will now reconsider the strain to which they are subjected, with a view to their adaptation to bear a given load with the least quantity of metal.

120. Suppose a beam supported at its ends, and bent by a weight laid at any intermediate point upon it: since all materials are both extensible and compressible, it is evident that the whole of the lower fibres are in some degree of extension, less or more, and the whole of the upper fibres are in a compressed state; there being some point, intermediate between both, where extension ends and compression begins. If then we suppose all the forces of extension and compression, in the section of the beam where the deflecting force is applied, to

be separately collected into two points, one over the other, the beam will offer the greatest resistance, the quantity of metal being the same, when these points are as far asunder as possible, since the leverage is then the greatest.

121. When the depth of the beam is limited, this object would, perhaps, be best attained by putting two strong ribs, one at the top and the other at the bottom, the intermediate part between the ribs being a thin sheet of metal, to keep the ribs always at the same distance, as well as to serve another purpose which will be mentioned further on.

122. As to the comparative strength of the ribs, in beams of different materials, that depends on the nature of the body, and can only be derived from experiment. Thus, suppose the same force were required to injure the elasticity to a certain extent, or to cause rupture, whether it acted by extension or compression, then the strengths of the ribs should be equal; and this would be the case whatever the thickness of the part between the ribs might be, providing it was constant. But, supposing the thickness of the part between the ribs was so small that its resistance might be neglected, and the metal to be of such a nature that a force F was needed to injure the elasticity to a certain degree by stretching it, and another force G to do the same by compressing it, it is evident that the size of the ribs should be as G to F, or inversely as their resisting power, that they may be equally affected by the strain. Or if, the resistance of the part between the ribs being neglected,

it took equal weights F′ and G′ to break the material by tension and compression, the beam should have ribs as G′ to F′ to bear the most without fracture.

123. This last matter must be considered with some modifications: it would not, perhaps, be proper to make the size of the ribs just in the ratio of the ultimate tensile to the crushing forces, as the top rib would be so slender that it would be in danger of being broken by accidents; and the part between the ribs, though thin, has some influence on the strength.

124. The thickness, too, of the middle part between the ribs is not a matter of choice: independent of the difficulty of casting, and the care necessary to prevent irregular cooling, and contraction, in beams whose parts differ much in thickness, the middle part cannot be rendered thin at pleasure, but must have a certain thickness, though in long beams the breaking weight is small, and a very small strength in the middle part is all that is necessary.

125. The neutral line being the boundary between two opposing forces, those of tension and compression, it seems probable that bending the beam would produce a tendency to separation at that place. Moreover, the tensile and compressive forces are, strictly speaking, not parallel; they are deflected from their parallelism by the action of the weight, which not only bends the beam, but tends to cut it across in the direction of the section of fracture; and this last tendency is resisted by all the particles in the section. This compounded force

will then tend to separate the compressed part of the beam, in the form of a wedge, and this tendency must be resisted by the strength of the part between the ribs or flanges. We have had several instances of fracture this way (arts. 86, 94), and there will occur several others in the course of the following experiments, as in art. 135, Experiment 12, &c.

126. We see then that there are three probable ways in which a beam may be broken: 1st, by tension, or tearing asunder the extended part; 2nd, by the separation of a wedge, as above; and 3rd, by compression, or the crushing of the compressed part. I have not, however, obtained a fracture, by this last mode, in cast iron broken transversely.

EXPERIMENTS TO ASCERTAIN THE BEST FORMS OF CAST IRON BEAMS, AND THE STRENGTH OF SUCH BEAMS.

127. In the commencement of these experiments the form I first adopted was one in which the arc, bounding the top of the beam, was a semi-ellipse, with the bottom rib a straight line; but the sizes of the ribs at top and bottom were in various proportions. The ribs in the model were first made equal, as in the beam of strongest form according to the opinion of Mr. Tredgold (Section IV., art. 37); and when a casting had been taken from it, a small portion was taken from the top rib, and attached to the edge of the bottom one, so as to make the ribs as one to two; and when another casting had been obtained, a portion more was taken

from the top, and attached to the bottom, as before, and a casting got from it, the ribs being then as one to four. In these alterations the only change was in the ratio of the ribs, the depth and every other dimension in the model remaining the same.

128. Finding that all these beams had been broken by the bottom rib being torn asunder, and that the strength by each change was increased, I had the bottom rib successively enlarged, the size of the top rib remaining the same. The bottom rib still giving way first, I had the top rib increased, feeling that it might be too small for the thickness of the middle part between the ribs. The bottom rib was again increased, so that the ratio of the strengths of the bottom and top ribs was greater than before; still the beam broke by the bottom rib failing first, as before. As the strength continued to be increased more than the area of the section, though the depth of the beam and the distance between the supports remained the same, I pursued, in the future experiments, the same course, increasing by small degrees the size of the ribs, particularly that of the bottom one, till such time as that rib became so large that its strength was as great as that of the top one; or a little greater, since the fracture took place by a wedge separating from the top part of the beam. I here discontinued the experiments of this class, conceiving that the beams last arrived at, were in form of section nearly the strongest for cast iron.

129. In most of the experiments the beams were intended to have been broken by a weight at their

middle; and, therefore, the form of the arcs, bounding the top of the beams, was, in this inquiry, of little importance: in making them elliptical, they were too strong near to the ends for a load uniformly laid over them; the proper form is something between the ellipse and the parabola. It is shown, by most of the writers on the strength of materials, that if the beam be of equal thickness throughout its depth, the curve should be an ellipse to enable it to support, with equal strength in every part, an uniform load; and if there be nothing but the rims, or the intermediate parts be taken away, the curve of equilibrium, for a weight uniformly laid over it, is a parabola. When, therefore, the middle part is not wholly taken away, the curve is between the ellipse and the parabola, and approaches more nearly to the latter, as the middle part is thinner.

130. The instrument used in the experiments was a lever (Plate II. fig. 40) about 15 feet long, placed horizontally, one end of which turned on a pivot in a wall, and the weights were hung near to the other; the beams being placed between them and the wall, at 2 or 3 feet distance from it.

131. All the beams in the first Table (Table I. following) were exactly $5\frac{1}{8}$ inches deep in the middle, and 5 feet long, and were supported on props just 4 feet 6 inches asunder. The lever was placed at the middle of the beam, and rested on a saddle, which was supported equally by the top of the beam and the bottom rib, and terminated in an arris at its top, where the lever was applied. The deflexions were taken in

inches and decimal parts, at or near the middle of the beam, as mentioned afterwards. The weights given are the whole pressure, both from the lever and the weights laid on, when reduced to the point of application on the beam. The dimensions of section in each experiment were obtained from a careful admeasurement of the beam itself, at the place of fracture, which was always very near (usually within half an inch of) the middle of the beam; the depth of the section being supposed to be that of the middle of the beam, or $5\frac{1}{8}$ inches.

132. As the experiments were made at different times, and there might be some variation in the iron, though it was intended always to be the same, a beam of the same length and depth as the others, but of the usual ⊥ form, always from the same model, was cast with each set of castings for the sake of comparison. The results of the experiments upon these beams are given in the third Table.

133. The first six beams, in the first Table, were cast *horizontal*, that is, each beam lay flat on its side in the sand; all the rest were cast *erect*, that is, each beam lay in the sand in the same posture as when it was afterwards loaded, except that the casting was turned upside down, when in the sand.

134. In all the experiments the area of the section was obtained with the greatest care; it includes, besides the parts of which the dimensions are given, the area of the small angular portions at the junction of the top and bottom ribs, with the vertical part between them.

TABLE

135. *Tabulated Results of experiments to ascertain the best form of ports 4 feet 6 inches asunder, and having the depth in the middle, are in inches, and the weights in pounds, except otherwise mentioned.*

Form of section of beam in middle.	Area of top rib in middle of beam.	Area of bottom rib in middle of beam.	Thickness of vertical part between the ribs.	Area of section of beam at place of fracture.	Weight of beam.	Deflexions in parts of an inch.	Corresponding weights in lbs.
1st Experiment. Beam (Plate III. fig. 41) with equal ribs at top and bottom. 	1·75 × ·42 =·735	1·77 × ·39 =·690	·29	2·82	36¼ lbs.		
2nd Experiment. Beam with area of section of top and bottom rib as 1 to 2. 	1·74 × ·26 =·45	1·78 × ·55 =·98	·30	2·87	39 lbs.		
3rd Experiment. Beam with area of section of top and bottom rib as 1 to 4. 	1·07 × ·30 =·32	2·1 × ·57 =1·2	·32	3·02	40 lbs.		
4th Experiment. Beam from the same model as the last, but cast the opposite way up.	1·05 × ·32 =0·34	2·15 × ·56 =1·20	·33	3·08	39½ lbs.		

I.

cast iron beams, all the beams being made 5 *feet long and laid on sup-*
where the load was applied, $=5\frac{1}{8}$ *inches. All dimensions in the Table*

Breaking weight in lbs.	Strength per square inch of section in lbs.	Strength per sq. inch of section of beam of usual form (Tab. III.) cast with these for comparison.	Gain in strength by comparison with beam of usual form.	Form of fracture.	Remarks.
6678 lbs. =59 cwt. 70 lbs.	$\frac{6678}{2\cdot82}$ =2368	2584	$-\frac{1}{12}$	This is represented by the line *b n r t*, (Plate III. fig. 41,) where $tr=\cdot6$, and $bn=2\cdot5$, the figure being a side view of the beam. The distances *t r*, *b n* are measured vertically.	
7368 lbs. =65 cwt. 88 lbs.	$\frac{7368}{2\cdot87}$ =2567	2584	$-\frac{1}{152}$	Nearly same as in Exp. 1; here *t r*, (Plate III. fig. 41,) =·55 inch. Here, and in all other cases, *t r* is measured vertically, as before.	
8270 lbs. =73 cwt. 94 lbs.	$\frac{8270}{3\cdot02}$ =2737	2584	$\frac{1}{17}$ nearly	Nearly as in Experiment 1; and *t r* =·6 (fig. 41).	
8263 lbs. =73 cwt. 89 lbs.	$\frac{8263}{3\cdot08}$ =2683	2792	$-\frac{1}{26}$ nearly	Nearly as figure to Exp. 1, but here $bn=2\cdot5$, and *t r* ·55 (fig. 41).	

TABLE I.—

Form of section of beam in middle.	Area of top rib in middle of beam.	Area of bottom rib in middle of beam.	Thickness of vertical part between the ribs.	Area of section of beam at place of fracture.	Weight of beam.	Deflexions in parts of an inch.	Corresponding weights in lbs.
5th Experiment. Beam with area of section of ribs as 1 to $4\frac{1}{2}$ nearly. 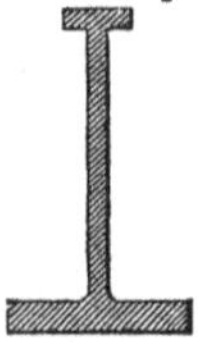	1·05 × ·34 =0·357	3·08 × ·51 =1·570	·305	3·37	$44\frac{3}{4}$ lbs.		
6th Experiment. Ratio of ribs 1 to 4 nearly. 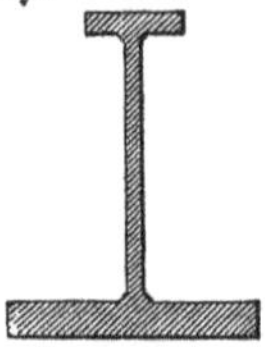	1·6 × ·315 =0·5	4·16 × ·53 =2·2	·38	4·50	57 lbs.	·4 ·45 ·52	11186 12698 13706
7th Experiment. Beam differing from last, having a broader bottom flange. Ratio of ribs 1 to $5\frac{1}{2}$ nearly. 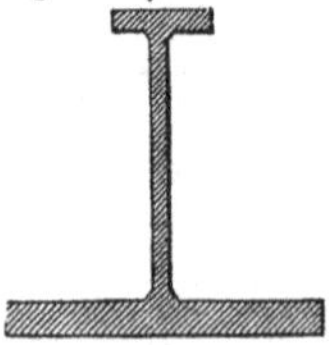	1·56 × ·315 =0·49	5·17 × ·56 =2·89	·34	5	$67\frac{1}{4}$ lbs.	·24 ·36 ·40 ·42 ·45 ·48 ·49 ·53	8288 12698 13706 14210 15218 15722 16226 16730
8th Experiment. 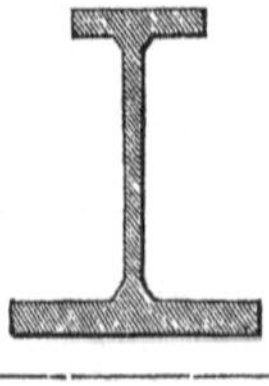	2·3 × ·315 =·72	4·06 × ·57 =2·314	·33	4·628			

(CONTINUED.)

Breaking weight in lbs.	Strength per square inch of section in lbs.	Strength per sq. inch of section of beam of usual form (Tab. III.) cast with these for comparison.	Gain in strength by comparison with beam of usual form.	Form of fracture.	Remarks.
10727 lbs. =95 cwt. 87 lbs.	$\frac{10727}{3 \cdot 37}$ =3183	2792	$\frac{1}{7}$ nearly	Here *t r* (fig. 41) =·6 inch.	Broke by tension, small flaw in bottom rib, at place of fracture.
14462 lbs. =129 cwt. 14 lbs.	$\frac{14462}{4 \cdot 5}$ =3214	2693	$\frac{1}{5}$ nearly	Here *b n* (fig. 41) =2·5 inches.	Broke by tension 1 inch from the middle.
16730 lbs. =149 cwt. 42 lbs.	$\frac{16730}{5}$ =3346	2693	$\frac{1}{4}$ nearly		After having borne the last-named weight some minutes, it broke by tension very near the middle.
15024 lbs. =134 cwt. 16 lbs.	$\frac{15024}{4 \cdot 628}$ =3246				Broke by tension very nearly in the middle. This beam and those in all the experiments, except the last, were of the form (Pl. III. figs. 42 and 43), being uniform in height, and having a large bottom rib tapering towards the ends.

TABLE I.—

Form of section of beam in middle.	Area of top rib in middle of beam.	Area of bottom rib in middle of beam.	Thickness of vertical part between the ribs.	Area of section of beam at place of fracture.	Weight of beam.	Deflexions in parts of an inch.	Corresponding weights in lbs.
9th Experiment. From the same model as that used in Experiment 8, except that the bottom rib is increased in breadth.	2·35 × ·29 = ·68	5·43 × ·537 = 2·916	·35	5·292		·12 ·15 ·18 ·20 ·22 ·25 ·26 ·29 ·31 ·33 ·53	6218 7598 8288 9309 10330 11338 12346 13354 14371 15393 16401
10th Experiment. Beam from the same model, but with further increase of bottom rib.		6·8 × ·502 = 3·413			64½ lbs.	·16 ·18 ·19 ·21 ·22 ·24 ·26 ·28	6218 7598 8288 9309 10331 11339 12341 13351
11th Experiment. Beam from same model as in last experiment.	2·3 × ·28 = ·64	6·61 × ·54 = 3·57	·34	5·86	68½ lbs.	·26 ·29 ·30 ·33 ·35 ·36 ·43	12087 12777 repeated 14345 15913 16697 18265
12th Experiment. 	2·33 × ·31 = ·72	6·67 × ·66 = 4·4	·266	6·4	71 lbs.		

(CONTINUED.)

Breaking weight in lbs.	Strength per square inch of section in lbs.	Strength per sq. inch of section of beam of usual form (Tab. III.) cast with these for comparison.	Gain in strength by comparison with beam of usual form.	Form of fracture.	Remarks.
16905 lbs. =150 cwt. 105 lbs.	$\frac{16905}{5\cdot292}$ =3194				Broke by tension.
14336 lbs. nearly =128 cwt.					This broke by tension, and ought to have borne considerably more than the last beam; but its iron must have been of a less tenacious kind than the others; as is evident by comparing their deflexions, this beam having bent little more than half what the preceding one did before it broke.
19441 lbs. =173½ cwt	$\frac{19441}{5\cdot86}$ =3317				This beam broke very nearly in the middle, by tension, as before.
26084 lbs. =11 tons 13 cwt.	$\frac{26084}{6\cdot4}$ =4075	2885	upwards of $\frac{2}{5}$	A wedge separated from its upper side, as shown in the fig below, which is a side view of the beam; where *adc* is the wedge, *ac*= 5·1 inches, *bd*= 3·9 inches, angle *adc* at vertex= 82°.	

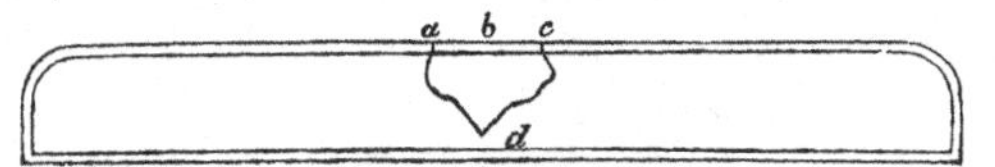

TABLE I.—

Form of section of beam in middle.	Area of top rib in middle of beam.	Area of bot tom rib in middle of beam.	Thickness of vertical part be- tween the ribs.	Area of section of beam at place of fracture	Weight of beam.	Deflex- ions in parts of an inch.	Corres- ponding weights in lbs.
13th Experiment. From the same model as in the last Experiment.	2·3 × ·28 =·64	6·63 × ·65 =4·31	·335	6·5	74¾ lbs.	·22 ·24 ·25 ·26 ·30 ·34 ·36 ·38 ·43 ·47 ·48 ·50	9328 11397 12777 14345 15913 17481 18265 19049 20617 22185 22969 "
14th Experiment. Elliptical beam, differing from that in Experiment 7, in having a larger bot- tom rib; ratio of ribs 6½ to 1.	1·54 × ·32 =·493	6·50 × ·51 =3·315	34	5 41	70¾ lbs.	·26 ·27 ·28 ·30 ·31 ·34 ·35 ·42 ·43 ·46 ·50 ·54	9327 10707 11397 12087 12777 14345 15913 16697 17481 19049 19833 20617

(CONTINUED.)

Breaking weight in lbs.	Strength per square inch of section in lbs.	Strength per sq. inch of section of beam of usual form (Tab. III.) cast with these for comparison.	Gain in strength by comparison with beam of usual form.	Form of fracture.	Remarks.
23249 lbs. =10 tons 8 cwt. nearly	$\frac{23249}{6\cdot5}$ =3576	2885	$\frac{1}{5}$ nearly.		Broke in the middle by tension.
21009 lbs. =9 tons 8 cwt. nearly	$\frac{21009}{5\cdot41}$ =3883	2885	upwards of $\frac{1}{4}$	Nearly as in figure to first Experiment, $bn = 1\cdot8$ inch.	

TABLE II.

136. *Results of experiments on beams whose forms (Plate III. figs. 42, 43) differ but little from the best of those in Table I.*

Form of section of beam in middle.	Distance between supports.	Depth of beam.	Area of top rib in middle of beam.	Area of bottom rib in middle.	Thickness of vertical part between the ribs.	Area of section at place of fracture.	Weight of beam.	Deflexions.	Corresponding weights.	Breaking weight.	Remarks.
Experiment 1. 	7 ft.	4·1 in.	2·25 × ·33 =·74	6·00 × ·74 =4·44	·40 in.	6·54	114 lbs.	·25	2764		This beam was cast 7 feet 6 in. long; and the "weight of beam" set down was in all cases that of the whole length.
								·26	2994		
								·27	3224		
								·29	3569		
								·32	3914		
								·40	5180		
								·44	5525		
								·45	6042		
								·55	6971		
								·62	7727		
								·70	8483		
								·75	8637		
								·80	10017		
								·95	11397		
								1·08	12815	13543 lbs.	
Experiment 2. 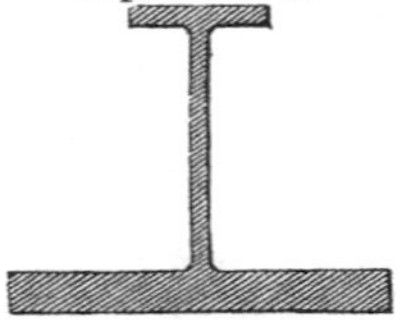	7 ft.	5·2 in.	2·25 × ·35 =·79	6·00 × ·77 =4·62	·34 in.	6·94	128 lbs.	·35	7947		Length of beam as before.
								·43	8637		
								·51	9327		
								·53	10017		
								·56	10707		
								·58	11397	15129 lbs.	
								·63	12087	=6 tons 15 cwt. 9 lbs.	

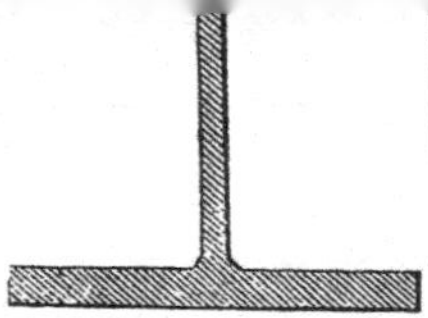								·36 ·42 ·49 ·56 ·58	10017 11397 13543 14999 15129	15129 lbs. =6 tons 15 cwt. 9 lbs.	
Experiment 4. 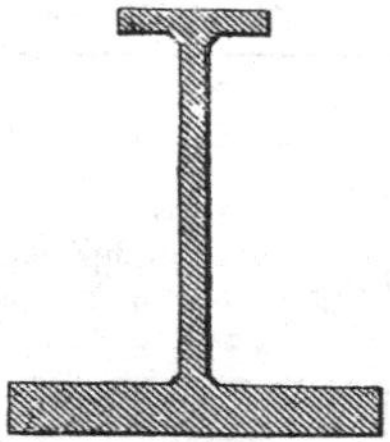	7 ft.	6·93 in.	2·25 × ·34 =·765	6·05 × ·75 =4·537	·38 in.	7·67	146 lbs.	·20 ·24 ·35 ·40 ·45 ·52 ·60 ·65	9327 10707 12087 14271 15913 17481 19049 20617 21401	22185 lbs. =9 tons 18 cwt.	
Experiment 5. Beam from the same model as the last.	7 ft.	6·98 in.	2·25 × ·32 =·72	5·95 × ·73 =4·343	·37 in.	7·40	141 lbs.	·25 ·36 ·44 ·47 ·50 ·58	9327 11397 14345 15913 17481 19049		
Experiment 6. 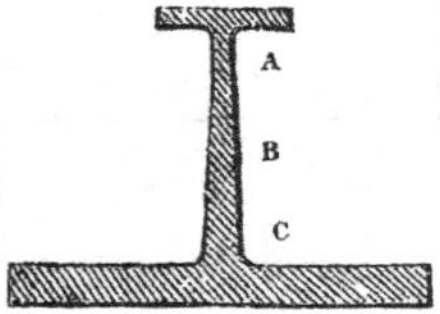 A B C	4 ft. 6 in.	$5\frac{1}{8}$ in.	2·15 × ·27 =·58	6·74 × ·71 =4·785	At A ·25 B ·37 C ·53	7·20	81 lbs.	·19 ·23 ·25 ·29 ·32 ·40 ·45 ·50 ·52 ·55	11056 12436 13816 15196 16576 19600 21616 23128 23632 24640	25144 lbs. =11 tons $4\frac{1}{2}$ cwt.	It is doubtful whether this beam broke by tension or compression, a crack showing a wedge which *broke out afterwards*, and of which *a b c d* is the form in figure below. *a c*=4·2 inches, *b d*=1·7 inches, or $\frac{1}{3}$rd of the depth of the beam nearly. The strength per square inch of section of this beam was 3492 lbs., and that of the beam of usual form in Table III., cast with it, was 2796 lbs.; hence gain in strength from form of section equal one-fifth nearly.

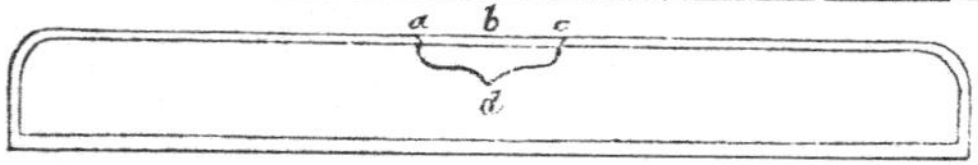

a b c
d

TABLE II.—(CONTINUED.)

Form of section of beam in middle.	Distance between supports.	Depth of beam.	Area of top rib in middle of beam.	Area of bottom rib in middle.	Thickness of vertical parts between the ribs.	Area of section at place of fracture.	Weight of beam.	Deflexions.	Corresponding weights.	Breaking weight.	Remarks.
Experiment 7. 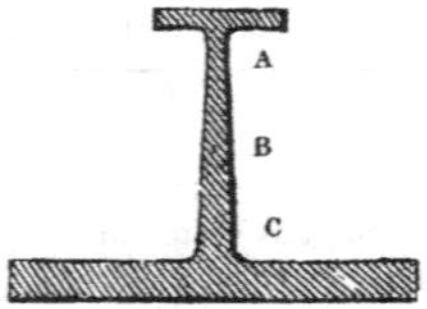	4 ft. 6 in.	5·1 in.	2·15 × ·24 = ·52	7·60 × ·72 = 5·472	At A ·27 B ·44 C ·48	7·90	88 lbs.	·17 ·21 ·25 ·27 ·28 ·32 ·35 ·37 ·42 ·52 ·55 ·58	10017 11397 12777 14345 15913 17481 18592 20608 22624 24640 26152 27664	28168 lbs. = 12 tons 11½ cwt.	This beam bore 3565 lbs. per square inch of section, and the beam of usual form 2796 lbs.; hence saving in metal from form of section = ·215.
Experiment 8. Beam of double length, and slightly varied section.	9 ft.	5⅛ in.	2·2 × ·36 = ·79	7·0 × ·69 = 4·83	A ·27 B ·33 C ·60		170½ lbs.	1·00 1·12 1·27 1·45	8296 8986 9676 10366	11056 lbs. = 4 tons 18¾ cwt.	Broke 9 inches from the middle, where there were two small defects whose area was about $\frac{1}{7}$th of an inch in the bottom rib. The experiment was therefore imperfect.
Experiment 9. Beam slightly differing from the last.	9 ft.	5⅛ in.	2·25 × ·3 = ·67	7·7 × ·76 = 5·85	A ·36 B ·42 C ·50		192 lbs.	·90 ·96 1·05 1·30 1·52 1·84 2·04	8296 8986 9676 11056 12436 13816 14506	15196 lbs. = 6 tons 15¾ cwt.	It broke in the middle, throwing out a wedge as in Experiment 6. Here *a c*, the length of the wedge, = 6·9 inches; *b d*, its depth, = 2·25 inches.

Experiment 10.	9 ft.	$10\frac{1}{4}$ in.	2·1 × ·27 = ·57	6·14 × ·77 = 4·72	At A ·26 B ·25 C ·35		227 lbs.	·23 ·26 ·34 ·40 ·46 ·51 ·55 ·61 ·65 ·68	11056 13816 16576 19600 21616 23632 25648 26656 27664 28168	28672 lbs. = 12 tons 16 cwt.	This beam broke as in Experiments 6 and 9; the dimensions of the wedge broken out were, *a c*, its length, = 13· inches *b d*, its depth, = 5·8 "
Experiment 11. Beam with top and bottom ribs somewhat larger than the last.	9 ft.	$10\frac{1}{4}$ in.	2·2 × ·33 = ·73	7·6 × ·75 = 5·70	A ·15 B ·38 C ·35		244 lbs.	·22 ·24 ·32 ·35 ·40 ·47 ·50 ·55 ·64 ·70 ·76	12436 13816 18592 20608 22624 24640 26656 28672 30184 31192 31696	32200 lbs. = 14 tons $7\frac{1}{2}$ cwt.	This broke as before, length of the wedge 18 inches, depth 6·15 inches. The form of the wedge was not so regular as before.

TABLE III.

137. *Results of Experiments upon beams of the usual form (Plate III. fig. 44, and section below), one of which was cast with each casting of those in Table I. Art.* 135, *for comparison with them. These beams were from the model of Messrs. Fairbairn and Lillie, in* 1830; *they were cast* 5 *feet long, and* $5\frac{1}{8}$ *inches deep in the middle, and were laid on supports* 4 *feet* 6 *inches asunder, as in the beams of Table I.*

	Thickness at A.	Thickness at B.	Thickness at C.	Thickness D E.	Breadth F E.	Area of section.	Weight of beam in lbs.	Deflexions.	Corresponding weights.	Breaking weight.	Strength per sq. inch of section in middle.
Experiment 1. 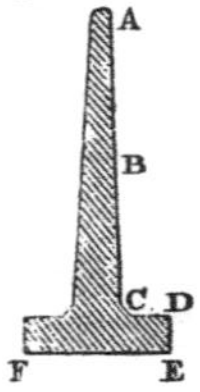	·32	·44	·47	·52	2·27	3·2	$40\frac{1}{2}$	·25 ·37	5758 7138	8270	$\frac{8270}{3\cdot2}=$ 2584 lbs.
Experiment 2.	·30	·37	·425	·53	2·28	2·98	38	·37 ·50 ·62	6679 9495 9279	9503	$\frac{9503}{2\cdot98}=$ 3188 lbs.
Experiment 3.	·29	·425	·46	·53	2·3	3·16	$40\frac{1}{2}$			8823	$\frac{8823}{3\cdot16}=$ 2792 lbs.
Experiment 4.	·29	·425	·53	·565	2·34	3·32	41	·4 ·43 ·47	7598 8494 8942	8942	$\frac{8942}{3\cdot32}=$ 2693 lbs.
Experiment 5.						3·08 nearly.	$39\frac{1}{2}$	·28 ·33	6218 7138	7598	$\frac{7598}{3\cdot08}=$ 2466 lbs.
Experiment 6.	·30	·42	·45	·51	2·28	3·17	40			9146 nearly.	$\frac{9146}{3\cdot17}=$ 2885 lbs.

Experiment 7.	·27	·40	·44	·46	2·27	2·921	36½	·21 ·23 ·24 ·26 ·27 ·29 ·31 ·33 ·35 ·37 ·38 ·40 ·42 ·44	4148 4493 4838 5183 5528 5873 6218 6563 6908 7253 7598 7943 8288 8540	8792 = 3 tons 18½ cwt.	$\frac{8792}{2·921}$ = 3009 lbs. The beam in Experiment 1, cast on its side like this, gave 2584 lbs. per sq. inch; hence the mean from the two is 2796 lbs.
Experiment 8.	·27	·355	·43	2·26	·47	2·837	37	·09 ·18 ·20 ·23 ·25 ·27 ·29 ·32 ·35 ·38 ·39 ·42 ·44 ·45 ·46	2078 4148 4493 4838 5183 5528 5873 6218 6563 6908 7253 7943 8288 8540 8792	9044 = 4 tons ¾ cwt.	$\frac{9044}{2·837}$ = 3188 lbs. Mean from the whole, 2851 lbs.

Some of these beams, particularly the first two, twisted in a serpentine manner before fracture.

REMARKS UPON THE EXPERIMENTS, IN TABLE I., TO OBTAIN THE BEST FORM OF CAST IRON BEAMS, OF WHICH THE LENGTH AND DEPTH WERE INVARIABLE.

138. It has been mentioned before, that these experiments were begun with that form of section of beam which Tredgold (Part I. art. 40) was induced to consider as the strongest, the top and bottom flanges in it being equal. This form was, however, found to be $\frac{1}{12}$th weaker to resist fracture, than that in common use (the ⊥ form); though it would, perhaps, be nearly the strongest in wrought iron. The top flange, in the model before used, was next reduced, and the part taken off it was added to the bottom one This alteration gave an increase of strength, and the beam, in Experiment 3, was somewhat stronger than that of the usual form cast with it for comparison. It did not now seem advisable to decrease further the top flange; and as every beam had been found to break by tension, or through the weakness of the bottom part, I thought it best to keep increasing the bottom flange, by small degrees, till such time as the beam broke by the rupture of some other part. Proceeding thus, the beam, in Experiment 5, had its bottom and top ribs, or flanges, as $4\frac{1}{2}$ to 1; and the result from that form of section was a gain in strength of about $\frac{1}{7}$th. Before increasing the bottom rib any further, I added a little to the top one, as the vertical part of the

beam, or that between the ribs, would be perhaps strong enough for much larger ribs In Experiments 6, 7, and 14, the top rib and vertical part of the model were the same, the only difference being in the increasing breadth of the bottom rib. From these, the gain in strength, above what was borne by beams of the usual form, was respectively $\frac{1}{5}$th, $\frac{1}{4}$th, and between $\frac{1}{4}$th and $\frac{1}{3}$rd.

In Experiments 8, 9, 10, 11, 12, 13, the top rib of the model was the same; but it was somewhat larger than in Experiments 6, 7, and 14, and the bottom rib was the only part intended to be varied. In the 12th experiment of the Table, (being the 19th made,) the section of the bottom rib at the place of fracture, the middle, was more than double the rest of the section there; and the ratio of the top and bottom ribs was as 1 to 6. In all the beams before this, fracture had taken place through the weakness of the bottom rib In this it took place by the separation of a wedge from the top rib and the vertical part of the beam, which part happened to be thinner than usual: the gain in strength, arising from the form of the section in this beam, was upwards of $\frac{2}{5}$ths of what the beam of usual form bore; and the saving from the general form of the beam was nearly $\frac{3}{10}$ths of the metal. This experiment was repeated, in Experiment 13, but the beam, though cast from the same model, had its vertical part thicker than in the former; and its strength, per square inch of section, was less. Fracture took place in it by the rupture of the bottom rib, as in all the preceding experiments except the last.

The form of section in Experiment 12 is somewhat better than that in Experiment 14; it is the best which was arrived at for the beam to bear an ultimate strain; and it is, doubtless, nearly the strongest which can be attained: for the vertical part between the flanges is as thin as it can, probably, be cast; and the remainder of the metal is disposed in the flanges, and consequently as far asunder as possible, the section of the flanges being in the ratio of 6 to 1, or nearly in that of the mean crushing and tensile strength of cast iron, (Part II. art. 33.)

The strengths, per square inch, borne by the beams in Experiments 12, 13, and 14, were 4075, 3576, 3883 lbs. respectively, the mean being 3845 lbs. And the strength of the beam of the common form, (Table III.,) cast with them for comparison, was 2885 lbs. per square inch. The difference, or strength gained from a mean among the results, was therefore 960 lbs., or upwards of $\frac{1}{4}$th of that mean.

Some of the beams, it will be noticed, had their bottom rib considerably thicker than the vertical part between the ribs; the line of junction being tapered from the thick to the thinner part. This tapering was more gradual, and higher up the beam, than is represented in the forms of section. The castings obtained were very good, as might be inferred from the strength of them; but as additional care is requisite to obtain good castings, when the parts differ much in thickness, we should bear in mind that it is not absolutely necessary, but convenient, to make one part thicker than another. The same strength would have been obtained by making the bottom rib broader and thinner than that in the beams tried, leaving the quantity of metal the same.

FORM OF BEAM ALTERED.

139. After Experiment 7, it was evident that the bottom flange in the future beams, if made of equal size throughout, as heretofore, would become very heavy; I had, therefore, the form of the beams, near to the ends, changed, leaving the section in the middle, as before, and making the height of the beam equal throughout, instead of being elliptical, as in the previous experiments. The bottom flanges were both made to taper toward the ends in the form of a double parabola whose vertex was in the middle of the beam; and the ratio of the sections of the flanges were, throughout the length of the beam, the same as that in the middle. (See Plate III. figs. 42 and 43, the former being a plan, and the latter an elevation of the beam.)

140. From the great quantity of matter in the flanges, particularly the bottom one, in the subsequent experiments, and the small thickness of the part between the flanges, I was convinced that nearly

all the tensile force would be exerted by the bottom flange, whilst the rest of the beam would serve for little more than a fulcrum; the centre of resistance to compression, or of that fulcrum, being very near to the top; it being perhaps at the point *r*, in the experiments in the Table, (Table I.)

Suppose then D to be the vertical distance from the centre of compression, at any part of the beam, to the centre of tension in the bottom flange; and if T be the direct tensile strength of the bottom flange at that part, T multiplied by some function of D, (or T D nearly,) will represent the strength of the beam there. But D throughout the same beam will be a constant quantity, or nearly so; the strength of the beam at any part, will, therefore, be nearly in proportion to that of its bottom flange at that part; and as the strain will be less toward the ends than elsewhere, the bottom flange will be reduced there likewise. Suppose the bottom flange to be formed of two equal parabolas, the vertex of one of them, ACB, being at C, in the figure annexed; then, by

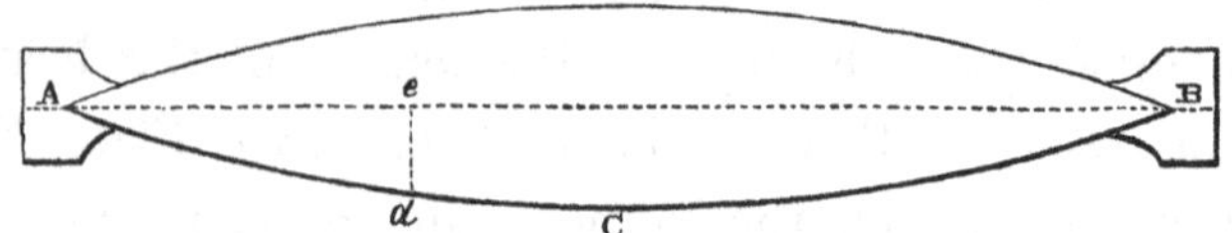

the nature of the curve, any ordinate *d e* is as A *e* × B *e*; the strength of the bottom flange, therefore, and consequently that of the beam at that place, will be as this rectangle. It is shown too, in Part I., and by writers on the strength of materials generally, that the rectangle $Ae \times Be$ is the proportion of strength which a beam ought to have, at any distance Ae from A, to bear equally the same weight in every part, or a weight laid uniformly over it. The conclusions above were verified by several experiments, in which beams were broken by weights applied at half the distance between the middle and the ends.

141. From the experiments in Table I., in which the length and depth of the beam were always the same, it would appear, that when the size of the top flange and the thickness of the vertical part remain unaltered, the latter being small, the strength of the beam is nearly in proportion to the size of the bottom flange; double the size of that flange giving nearly, but not quite, double strength.

REMARKS ON TABLE II.

142. The first four beams in this Table were all cast 7 feet 6 inches long, and they were supported by props 7 feet asunder. They were all from the same model, which varied only in the breadth of its vertical part between the flanges, the depth of the beam being all that was intended to vary. The depths were nearly 4, 5, 6, and 7 inches, but accurate admeasurements are given with the sections. The vertical part in these beams was rendered too strong comparatively with the size of the bottom flange, in order that they might all break by tension,

or in the same manner, to furnish the means of judging correctly of their relative strength.

143. From these experiments it appears that the ultimate strength, in sections like the preceding, is, *cæteris paribus*, nearly as the depth, but somewhat lower than in that ratio.

SIMPLE RULE FOR THE STRENGTH OF BEAMS APPROACHING TO THE BEST FORM IN THE PRECEDING EXPERIMENTS. (TABLES I. AND II.)

144. It appears, from art. 141, that when the length, depth, and top flange, in different cast iron beams, with very large flanges, are the same, and the thickness of the vertical part between the flanges is small and invariable, the strength is nearly in proportion to the size of the bottom flange. It appears, too, from the last article, that in beams which vary only in depth, every other dimension being the same, the strength is nearly as the depth.

145. Hence in different beams whose length is the same, the strength must be nearly as their depth multiplied by the area of a middle section of their bottom flange; and when the length is different, the strength will be as this product divided by the length.

$$\therefore \mathrm{W} = \frac{c\,a\,d}{l},$$

where W = the breaking weight in the middle of the beam, a = the area of a section of the bottom flange in the middle, d = the depth of the beam there, l = the distance between the supports, and c

= a quantity nearly constant in the best forms of beams, and which will be supplied from the results of the experiments in Tables I and II.

146. We will seek, by means of this approximate formula, for the value of c considered as constant, obtaining it from each of the experiments; and, for that purpose, confining ourselves to those forms in which the section of the bottom flange in its middle is more than half the whole section of the beam, take the mean from among them all for c.

$$\text{Since } W = \frac{c\,a\,d}{l}, \therefore c = \frac{l\,W}{a\,d}.$$

Taking the dimensions in inches, and the breaking weight in cwts., and separating the results of the beams which were cast *erect* from those cast *on their side*, we shall have

In beams cast erect.			*In beams cast on their side.*		
Experiment		Value of c.	Experiment		Value of c.
7	Table I.	545	6	Table II.	494
9	„	545	7	„	484
11	„	512	9	„	489
12	„	558	10	„	571
13	„	507	11	„	531
14	„	596		Mean	514 cwt.
1	Table II.	558			
2	„	472			
4	„	529			
	Mean	536 cwt.			

Since 536 or 514 is the mean value of c to give the breaking weight in cwts., according as the beam has been cast erect or on its sides, one-twentieth of these numbers, or 26 8 and 25 7, will be the value of c to give the breaking weight in tons. Neglecting the decimals, and taking 26 for the mean value of c, we have

$$W = \frac{c\,a\,d}{l} = \frac{26\,a\,d}{l}$$

for the strength in tons where the dimensions d and l are in inches.

But if l, the distance between the supports, be taken in feet, the value of c will be $\frac{26}{12} = 2{\cdot}166$, and the strength in tons will be

$$W = \frac{2{\cdot}166\,a\,d}{l}.$$

This rule is formed on the supposition, that the strength of the flanges is so great that the resistance of the middle part between them is small in comparison, and may be neglected.

Another approximate rule, for the strength of the beams in Tables I. and II., and which includes the effect of the vertical part between the flanges, may be deduced as below.

147. Since cast iron resists rupture by compression with about $6\frac{1}{2}$ times the power that it does by extension, (art. 33,) we may consider it as comparatively incompressible, and suppose that the operation of the top flange of the beam, when bent, is only to form a fulcrum upon which to break the bottom flange and the part between the flanges.

Let then A D E represent the section of the beam in the middle, D E being its bottom flange and A the top one, round which it turns.

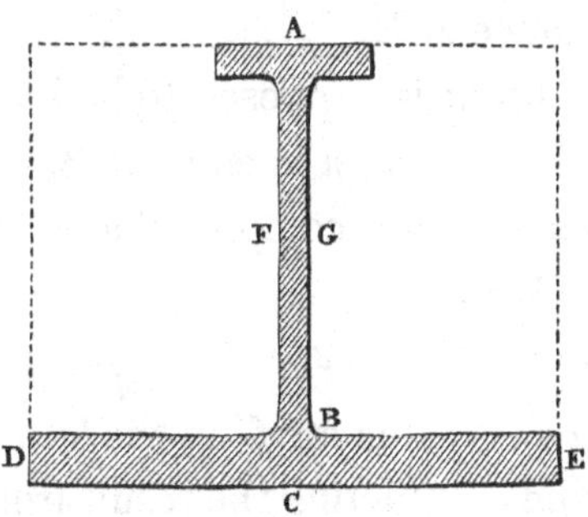

Let W = breaking weight.

l = distance between supports.

d = A C = the whole depth.

d' = A B = the depth to the bottom flange.

b = D E = the breadth of the bottom flange.

b' = F G = the thickness of the vertical part.

f = the tensile strength of the metal per unit of section.

n = a constant quantity.

To seek for the strength of the beam we may estimate, first, the resistance of a rectangular solid whose depth is A C and breadth D E, as shown by the dotted lines, and subtract from that the resistance which would be offered by that part which the beam wanted to make it such an uniform solid as the above.

$\therefore \frac{f b d^2}{n}$ = moment of resistance of the particles in the rectangular solid A D E, and $\frac{f'(b-b')d'^2}{n}$ = moment of resistance of the part necessary to make the beam a solid rectangle, where f' is the strain of the particles at the distance A B.

When the beam is supposed to be incompressible, as in the present case, n is equal to 3, and when it is equally extensible and compressible, n is equal to 6, (Tredgold, art. 110.)

$$\text{But } f' : f :: d' : d, \therefore f' = \frac{fd'}{d}.$$

Substituting this value for f' in the latter moment of resistance, and subtracting the result from the former moment, gives the moment of resistance of the solid equal to

$$\frac{fbd^2}{n} - \frac{fd'}{d} \times \frac{(b-b')d'^2}{n}.$$

But, from the property of the lever, this moment is equal to $\frac{1}{2}l \times \frac{W}{2} = \frac{lW}{4}$, or to half the weight acting with a leverage of half the length.

$$\text{Whence } \frac{lW}{4} = \frac{fbd^2}{n} - \frac{fd'}{d} \times \frac{(b-b')d'^2}{n}$$

$$= \frac{f}{nd}\left\{bd^3 - (b-b')d'^3\right\}$$

$$\therefore W = \frac{4f}{nld}\left\{bd^3 - (b-b')d'^3\right\}.$$

In the same iron, f and n are constants; putting $c = \frac{4f}{n}$, we shall have

$$W = \frac{c}{dl}\left\{bd^3 - (b-b')d'^3\right\}.$$

This formula gives

$$c = \frac{dlW}{bd^3 - (b-b')d'^3}.$$

The numerical value of c, calculated by this formula, from each of the experiments in Tables I. and II., taking the breaking weight in ℔s., the length in feet, and the other dimensions in inches, is as below.

	Value of c.		Value of c.
Table I.	1897·98	Table II.	1932·78
	1735·94		1566·36
	1666·31		1593·82
	1638·92		1607·46
	1744·10		1604·63
	1725·90		1623·98
	1631·74		1627·43
	1785·96		1547·49
	1570·80		
	1835·14		
	1671·01		
	1797·87		

Mean value of c from the whole twenty beams, 1690·28.

This value of c is in ℔s., and dividing it by 2240, gives ·7544 for its value in tons.

The iron in my experiments on beams was of a strong kind, made with a cold blast; and many of the beams were cast erect in the sand, which gives them a little additional strength. We may, therefore, expect that the value of c, just obtained, will be somewhat too great for the generality of hot-blast castings; and for large beams, the iron of which is usually softer than that of small ones. We will, therefore, collect here its values, to obtain the strength in tons, computed from the results of other experiments on a large scale given further on. Taking them in the order in which they are inserted, we have as follows:

From Messrs. Marshall's beams (art. 153), cast from the cupola, we obtain	$c = \cdot 625$	Mean $\cdot 671$
"	$c = \cdot 710$	
From Mr. Gooch's beam (art. 154), we obtain	$c = \cdot 679$	
Messrs. Marshall's beam (art. 153), cast from the air furnace, gives	$c = \cdot 795$	

Mr. Cubitt's beams (art. 165), taking the results from the sound ones only, and the value of b' from a mean where that dimension varies, give as below.

From Experiment 1	$c = \cdot 6523$	Mean $\cdot 7086$
" " 2	$c = \cdot 6467$	
" " 3	$c = \cdot 7411$	
" " 4	$c = \cdot 7276$	
" " 6	$c = \cdot 7473$	
" " 9	$c = \cdot 6703$	
" " 10	$c = \cdot 7746$	

Taking a mean value of c, as obtained from the whole of the beams cast from the cupola, thirty in number, we should have it considerably more than ·7; the means from twenty experiments being ·7544; from three experiments ·671; and from seven experiments ·7086. The lowest of these means differs but little from $\frac{2}{3}$; and adopting this as a safe approximate value for c, from which to compute the strength of beams generally, we have in the preceding formula for W its value as below.

$$W = \frac{2}{3\,d\,l}\left\{b\,d^3 - (b - b')\,d'^3\right\},$$

where W is in tons, l in feet, and the rest are in inches.

148. The preceding formula for the strength of a beam depends on the two following suppositions: 1st, that all the particles, except those of the top part or flange of a bent beam, are in a state of tension;

2nd, that the resistance of each particle is as its distance from the top of the beam. Neither of these suppositions can be regarded otherwise than as an approximation. We know that the former, which is almost tantamount to the exploded assumption of Galileo, that materials are incompressible, is not strictly true of any bodies whatever; and the 2nd supposition is subject to the double inaccuracy of the leverage of the particles being estimated as from the top of the beams, and therefore rather too great; and of the force of the fibres being as their extension, whilst, in reality, it is in a less ratio than that, as shown in preceding articles (99 to 107).

If, as is expected, the formula should be allowed to give results agreeing moderately well with those of experiment at the time of fracture, it will appear evident that the 2nd assumption above is favourable to that of incompressibility, in estimating the transverse strength of cast iron.

149. To obtain further evidence on this subject, we will seek, by means of the experiments in this work, for the value of n in the formula, $w=\frac{f b d^2}{n l}$, for the strength of a rectangular bar, fixed at one end and loaded at the other, b being the breadth, d the depth, l the length of leverage, w the weight at the end, and the rest as in art. 147.

150. Selecting from the experiments, in articles 3 and 96, the mean tensile and transverse strengths of all the irons in which both these properties were obtained, we have as in the following Table.

Description of Iron.	Tensile strength per square inch. (f)	Transverse strength of bar 1 inch square and 54 inches between supports.	Transverse strength of bar 1 inch square, fixed at one end and loaded at the other, the weight acting with with a leverage l of 27 inches. (w)	Value of n from formula $n = \frac{f b d^2}{l w}$.
	lbs.	lbs.	lbs.	
Carron iron, No. 2, cold blast . .	16,683	476	238	$n = 2·59$
Carron iron, No. 2, hot blast . .	13,505	463	231½	$n = 2·16$
Carron iron, No. 3, cold blast . .	14,200	446	223	$n = 2·36$
Carron iron, No. 3, hot blast . .	17,755	527	263½	$n = 2·50$
Devon iron, No. 3, hot blast . .	21,907	537	268½	$n = 3·02$
Buffery iron, No. 1, cold blast .	17,466	463	231½	$n = 2·79$
Buffery iron, No. 1, hot blast . .	13,434	436	218	$n = 2·28$
Coed-Talon iron, No. 2, cold blast	18,855	413	206½	$n = 3·38$
Coed-Talon iron, No. 2, hot blast	16,676	416	208	$n = 2·96$
Low Moor iron, No. 3, cold blast	14,535	467	233½	$n = 2·30$
				Mean value of $n = 2·63$

151. The transverse strength of a rectangular body being directly as the product of the breadth, the square of the depth, and the strength of its fibres, and inversely as the length, the value of w, in the formula $w = \frac{f b d^2}{n l}$, will depend upon the value of n; and this last quantity will, as we have seen, depend on the comparative resistance of the fibres to extension and compression. Thus if, according to the general assumption, the extensions and compressions of the particles are equal from equal forces, and as the forces, the neutral line will be in the middle of the body, and the value of n equal to 6 (Part I. art. 110). If, according to Galileo, the body were incompressible, and the forces of the fibres were as their extension, we should have $n = 3$; and if, on the supposition of incompressibility, the forces of the fibres were the same for all degrees of extension, we should have $n = 2$. (See my paper on the Strength of Materials, 'Memoirs of the Literary and Philosophical Society of Manchester,' vol. iv. 2nd series, p. 243.)

The value of n, in the preceding Table, obtained from numerous experiments upon ten kinds of cast iron, varies from 2·16 to 3·38, the mean being 2·63. This result shows that the assumption of the incompressibility of cast iron may be admitted so long as we assume that the forces are directly as the extension of the fibres; and it might be admitted still, if we were to make the more improbable assumption, that the forces are the same for all degrees of extension; for the value of n in the former case would be 3, and in the latter 2, and the mean result is 2·63, somewhat nearer to the former than the latter.

The mean value of n obtained from the fracture of different kinds of stone, in numerous experiments not yet published, is not widely different from 3. The value of n, being assumed by Tredgold as 6, has, when applied at the time of fracture, caused the errors pointed out in notes to arts. 68, 143, &c. of Part I.

152. To obviate the anomalies above, and to obtain results consistent with experiment in the fracture of beams of cast iron—taking the neutral line in its proper position—we shall assume the forces of extension and compression to be of the form $f = a x - \phi(x)$, where f is the force, x the extension or compression, a a constant quantity, and $\phi(x)$ a function representing the diminution of the force f in consequence of the defect of elasticity. If $\phi(x)$ be assumed as equal to $b x^n$, as in art. 106, we shall have $n = 2$ nearly, if the experiments are made on the transverse flexure of bars; but it is more desirable that the value of n should be obtained from the direct longitudinal variations of the body experimented upon. This subject will be resumed in a future article.

EXPERIMENTS ON LARGE BEAMS.

153. I have been favoured, through Mr. J. O. March, of Leeds, with the results of three experi-

ments upon beams cast for the mill of Messrs. Marshall and Co., of that town, in 1838. They were from drawings supplied by Mr. Fairbairn, of Manchester; and were of a moderately good form of section, according to my experiments, though the bottom flange was rather too small. The beams were broken to ascertain the ultimate strengths, as well as to test the difference of strength between those cast from the cupola and the air furnace. The experiments, Mr. March states, were very carefully made, under the inspection of Messrs. Marshall and Co.; and the beams were cast from Bierley pig iron. The form, dimensions, and results are as follow:

Dimensions of the beams.

12 inches deep at one end, and
$10\frac{1}{2}$,, deep at the other.
$2\frac{1}{2}$,, = breadth of rib on the upper edge at the ends.
4 ,, = breadth of flange at the ends.

Dimensions of section in middle, in inches.

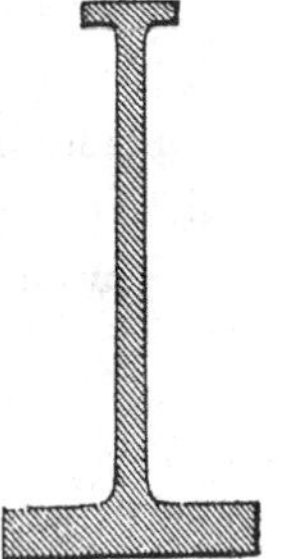

Area of top rib 3·00 × ·75 = 2·25.

Area of bottom rib 8·25 × 1·25 = 10·31, or 11 square inches, the increase from the brackets, at the junction of the bottom rib with the vertical part, being included.

Thickness of vertical part, $\frac{5}{8}$ = ·625 inch.

Depth of beam, in middle, 17 inches.

Beams proved at Messrs. Marshall and Co.'s, Leeds; the distance between the supports 18 *feet. Deflexions in* 50*ths of an inch.*

Tons.	2	4	6	8	10	12	14	16	18	20	22	24	25	26	27	28	
Deflected.	5	9	13	20	25	30	36	40	47	54	58						1st cupola casting. Broke with 22 tons.
Permanent Deflexion.				4	5	6	7	8	8	10							
Deflected.		7		12		26		36		47	54	61					2nd cupola casting. Broke with 25 tons.
Permanent Deflexion.				4		6		6		8	10	12½					
Deflected.		7		16		32		35		47	53	60	65	67	72		Air furnace casting. Broke with 28 tons.
Permanent Deflexion.				1		2		4		7	8	24	12	15	17½		

154. In an experiment made upon a beam by Mr. Gooch, whilst he was superintending the formation of the Leeds and Manchester Railway, the particulars are as follow:

Dimensions of section in middle, in inches.

Top rib . . 6 × 1½ = 9.
Bottom rib 8 × 1½ = 12.
Thickness of middle part 1½.
Depth of beam . . . 9.

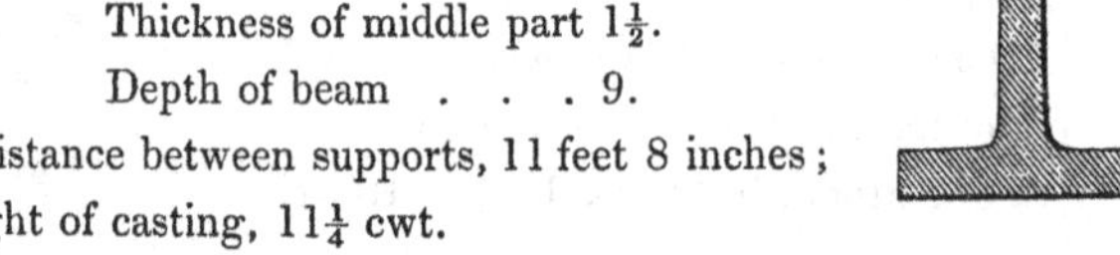

Distance between supports, 11 feet 8 inches; weight of casting, 11¼ cwt.

Weights laid on, in tons.	Deflexions, in parts of an inch.
3½	·10
4½	·15
5½	·175
6½	·22
7½	·25
8½	·30
10	·33
17	1·10
20	Broke 8½ inches from centre.

After bearing 17 tons, the beam was unloaded, and the elasticity seemed to be very little, or none, injured.

The mixture of metal was

1 ton Colebrook Vale, cold blast.
1½ „ Staffordshire, hot blast.
1½ „ Scotch, „

Mr. Gooch observes, in his letter giving an account of the experiment, "The cross section and some of the other dimensions are not of the most favourable or economical form, but circumstances required the adoption of them in the case of this girder."

155. The two following beams were cast for a viaduct forming the junction of the Liverpool and Manchester with the Leeds Railway, passing through Salford. They are not of forms best adapted for resisting fracture, but their great size will give additional interest to experiments upon them.

156. As several beams were cast from the same models, I was requested, by the Messrs. Ormerod, of Manchester, the founders, to superintend an experiment upon a beam from each; but the strength was so great that the experiment could not well be made in the usual way, that of applying a weight in the middle. I had, therefore, the beam inverted during the experiment, its bottom flange being turned upwards. The middle was supported laterally by stays, and it rested upon a cross bar, and other apparatus, on which it turned as on an axis; this cross bar being made to rest on two other

beams of nearly equal magnitude to that intended to be tried. One end of these beams, and the corresponding end of the beam to be bent, were connected by means of a strong bolt; and the other end of the latter beam was connected, as described below, with the opposite ends of the two supporting beams. The object was to break the top beam in the middle by a force applied at one end, whilst the other end was fixed; and to effect the required pressure, a powerful lever, forged for the purpose, was applied to the moveable end; it being evident that the pressure at the end would only be half of the effect produced in the middle. (See Plate V.)

157. To enable the deflexions to be observed, a straight edge of the same length as the beam was used, and made to rest upon it, touching it only at the ends. The quantities, which the deflexions varied in consequence of different weights, were measured by inserting a long wedge-like body, graduated on the side, between the straight edge and the top of the beam. The observed distances, when the beam was bearing a given load, were subtracted from the observed distance when there was no load upon it, for the deflexion.

The experiments were made with a very complete apparatus and every attention to accuracy.

158. The first beam had small ribs, or flanges, at top and bottom, and a strong vertical plate between them, as in the annexed section. The dimensions and results from the beam are as follow:

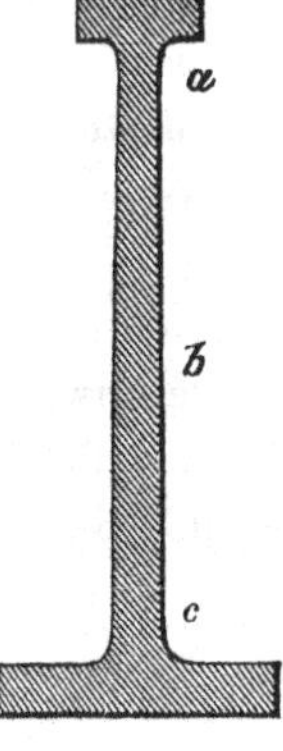

Section of top rib 5·1 × 2·33 inches.

" bottom rib 12·1 × 2·07 "

Thickness at *a*, 2·06 }
" *b*, 2·12 } mean 2·08.
" *c*, 2·07 }

Depth of beam in middle 30·5 inches.

Distance between supports 27 feet 5 inches.

Whole length of beam 28 feet 9 inches.

Weight applied at the end, in tons.	Weight applied in the middle, in tons.	Deflexion in middle, in parts of inch.
9·2	18·4	·31
13·4	26·8	·47
17·6	35·2	·61
21·8	43·6	·73
26·0	52·0	·87
30·2	60·4	1·02
34·4	68·8	1·16
38·3	76·6	1·29

With this, 76·6 tons in middle, it broke apparently in consequence of an accidental shake.

159. The second beam was much heavier and stronger than the former; its section is as in the annexed figure, and its dimensions and results are as follow.

Mean thickness of bottom flange 3·12 in.
Breadth of ,, 23·9 ,,
Whole depth of beam . . . 36·1 ,,
Thickness of vertical rib at *a* . 3·14 ,,
,, ,, *b* . 3·36 ,,
,, ,, *c* . 3·38 ,,
Distance between supports 23 feet 1 inch.
Whole length of beam 24 ,, 6 ,,
Weight of beam, 6 tons 1 cwt. 1 qr.

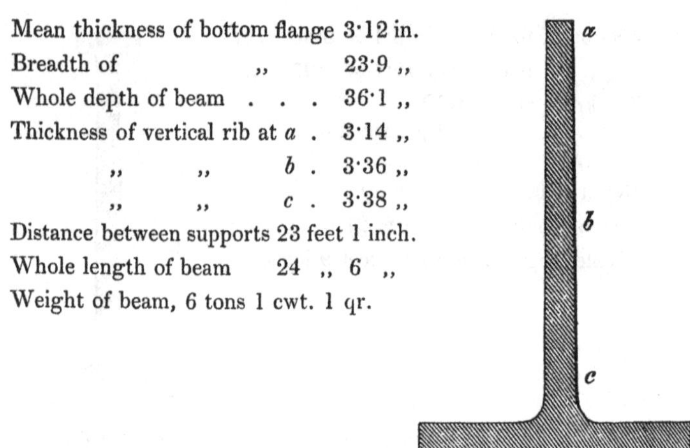

Weight applied at end of beam, in tons.	Weight applied at middle of beam, in tons.	Deflexion in middle of beam, in inches.
13·4	26·8	·10
17·6	35·2	·14
21·8	43·6	·17
26·0	52·0	·20
30·2	60·4	·23
34·4	68·8	·27
38·6	77·2	·31
42·8	85·6	·35
47·0	94·0	·38
51·2	102·4	·42
55·4	110·8	·46
59·6	119·2	·51
63·8	127·6	·55
68·0	136·0	·59
72·2	144·4	·64
76·4	152·8	·68

With this load, 153 tons in the middle nearly, the experiment was discontinued, as the apparatus was overstrained.

MR. FRANCIS BRAMAH'S EXPERIMENTS ON BEAMS.

160. In the second volume of the Institution of Civil Engineers, there is a paper containing experiments made in the year 1834, by Mr. A. H. Renton, for Mr. Bramah. They were upon beams of which the section is in the form ⊥, some of them being solid throughout their length, and others having apertures in them. The following is an abstract of such of the experiments as were pursued to the time of breaking the beam.

161. Beams, the section of which is of the ⊥ form, are, as we have seen, much weaker than those of another which has been arrived at (art. 135); but they are not without interest, as they are easily cast, and have considerable strength; except those with open work in them, which form the subject of Mr. Bramah's second Table, given hereafter; his first Table containing the results of experiments on solid beams. The former beams, though recommended by Tredgold (Part I. art. 41), are very weak; as will be seen by comparing the breaking weights of the beams, 3 inches deep, in the two Tables, and taking into consideration the weights of the beams. The great weakness of beams with apertures in them was shown, too, in my experiments on beams, 'Memoirs of the Literary and Philosophical Society of Manchester,' second series, vol. v. 1831.

TABLE I.

162. *Mr. Bramah's Experiments upon solid cast iron beams, the middle section of which was as in the annexed figure; the width of the flange throughout the beam being 1·5 inch, the thickness of the beam in every part ·5 inch, the depth of the section in the middle 3 inches, and the distance between the supports 3 feet 1 inch.*

Weight laid on scale with a leverage of 12.	Real weight laid on beam.	Beam 3 inches deep, and uniform in depth throughout, (Plate IV. fig. 45,) broke with flange downwards. ⊥ Weight 22 ℔s. 4 oz.	Beam same as the last, broke with the flange upwards. T Weight 21 ℔s. 2 oz.	Beam, with middle section as before, but reduced at the ends to half the depth of the middle, broke with flange downwards. ⊥ Weight 17 ℔s.	Beam, middle section as before, reduced at ends to $\frac{2}{3}$rds of depth in middle, broke with flange downwards. ⊥ Weight 17 ℔s. 12 oz.	Beam same as last, but reversed, broke with flange upwards. T Weight 17 ℔s. 8 oz.
℔s.	℔s.					
56	672	·027	·025	·037	·042	·036
112	1344	·0535	·049	·072	·078	·073
168	2016	·079	·075	·108	·113	·113
224	2688	·102	·105	·145	·144	·151
280	3360	·126	·132	·185	·181	·199
336	4032	·148	·166	·219	·220	
364	4368	·172	·183	·237	·240	
392	4704	·186		·257	·262	
420	5040	·205			·284	
448	5376				·306	
504	6048				·352	
532	6384				·393	
		144 × 12 = 1728 ℔s.* injured the elasticity; 434 × 12 = 5208 ℔s. broke the beam.	160 × 12 = 1920 ℔s. injured the elasticity; 378 × 12 = 4536 ℔s. broke the beam.	160 × 12 = 1920 ℔s. injured the elasticity; 511 × 12 = 6132 ℔s. broke the beam.	158 × 12 = 1896 ℔s. injured the elasticity; set with that weight ·004; 546 × 12 = 6552 ℔s. broke the beam.	156 × 12 = 1872 ℔s. injured the elasticity; 294 × 12 = 2528 ℔s. broke the beam. There was an air bubble in the casting.

* This must not be considered as the first weight which injured the elasticity, but that which rendered the defect very obvious. It has been shown (arts. 86 and 101) that the elasticity would be injured by any weight, however small. The same remark will apply to the other beams.

TABLE II.

163. *Beams differing in section from the former only in having a portion taken away from the middle; the sections being of the general form in the margin, and the elevations as in the Plates referred to below.*

Distance of supports = 3 feet 1 inch.

Depth of beam in middle, except otherwise mentioned, 3 inches.

Weight laid on scale with leverage of 12.	Real weight on beam.	Beam of form as in Plate IV. fig. 46, weight 15 lbs. 8 oz.; broke with the flange downwards.	Beam same as last, broke with flange upwards; weight 16 lbs. 8 oz.	Beam of form as in Plate IV. fig. 47, weight 17 lbs. 6 oz.; broke with flange downwards.	Beam same as last, weight 18 lbs. 6 oz.; broke with flange upwards.	Beam of form as in Plate IV. fig. 48, weight 15 lbs. 7 oz.; broke with flange downwards.	Beam same as last, broke with flange upwards.	Beam same as in Plate IV. fig. 49, depth in middle 4 in., weight 17 lbs. 2 oz.; broke with flange downwards.	Beam same as last, broke with flange upwards.	Beam same as in Plate IV. fig. 50, part taken out of centre, same area of section in middle as last, weight 21 lbs. 8 oz.; broke with flange downwards.	Beam same as last, broke with flange upwards.
		Deflexion.	Deflexion.	Deflexion.	Deflexion.	Deflexion.	Deflexion.	Deflexion.	Deflexion.	Deflexion.	Deflexion.
lbs.	lbs.	inch.	inch.	inch.	inch.	inch.	inch.	inch.	inch.	inch.	inch.
28	336	·050	·048	·017	·017	·023	·026	·012	·010	·009	·009
56	672	·099	·100	·037	·035	·048	·053	·027	·024	·019	·019
84	1008	·152	·155	·055	·052	·070	·081	·039	·037	·029	·030
98	1176	·182	·190	....	....	....	....	....	....	....	....
112	1344	....	....	·075*	·070*	·091	·109	·051	·050	·037	·031
140	1680	....	....	·092	·092	·113	·135	·062	·061	·044	·042
168	2016	....	....	·113	·115	·139	·164	·075*	·075*	·051	·052
180	2160	....	....	....	·122	....	·176	....	....	....	....
196	2352	....	....	·134	....	·165	....	·088	·086	·059	....
224	2688	....	....	·156	....	·190	....	·101	·099	·069	....
252	3024	....	....	·179	....	·218	....	·111	·111	·079	....
280	3360	....	....	·202	....	·246	....	·121	....	·090	....
308	3696	....	....	·224	....	....	....	·133	....	·099	....
336	4032	....	....	·245	....	....	....	·147	....	·110	....
392	4707	....	....	·292	....	....	....	·172	....	·133	....
448	5376	....	....	....	....	....	....	·200	....	·159	....
504	6048	....	....	....	....	....	....	·228	....	·186	....
532	6384	....	....	....	....	....	....	·242	....	·200	....
560	6720	....	....	....	....	....	....	·258	....	....	....
588	7056	....	....	....	....	....	....	·271	....	....	....
		Broke by 100 × 12 = 1200 lbs.	Broke by 98 × 12 = 1176 lbs.	Broke by 400 × 12 = 4800 lbs.	Broke by 184 × 12 = 2208 lbs.	Broke by 284 × 12 = 3408 lbs.	Broke by 188 × 12 = 2256 lbs.	Broke by 592 × 12 = 7104 lbs.	Broke by 276 × 12 = 3312 lbs.	Broke by 540 × 12 = 6480 lbs.	Broke by 196 × 12 = 2352 lbs.

* The asterisks show the deflexions with which the elasticity was observed to be injured.—See note to last Table.

164. Mr. Bramah's experiments were made to support certain principles adopted by Tredgold, some of which have been controverted in this work. Mr. Bramah states that it was "a principal feature in these experiments, and essential to the accuracy of the results, to note that point where the elastic power becomes impaired, and the specimens take a permanent set," &c. But I have shown (arts. 86 and 101) that there is no elastic limit in cast iron; and if there were, the depth, 3 inches, of the beams on which Mr. Bramah's experiments were made, was much too great, with their small distance between the supports, 3·083 feet, to enable him to discover when the defect of elasticity first took place. In neglecting his reasonings, as very disputable, and not suited to this place—where the strength considered is the ultimate, and Bramah's (following Tredgold) the incipient, of the material—I shall merely observe, that the results, with respect to fracture, seem to be in accordance with those from my own experiments (arts. 112-115). Mr. Bramah draws no conclusions from the breaking weights, but they show the great difference in the ultimate strength of a beam, according as it is turned one way up, or the opposite; and will serve as a beacon to warn the public of the danger of using beams with apertures in them.

MR. CUBITT'S EXPERIMENTS ON BEAMS.

165. Sir Henry De la Beche and Thomas Cubitt,

Esq., having in 1844 been appointed by Government to inquire into the circumstances respecting the fall of a cotton mill at Oldham, in Lancashire, and part of a prison at Northleach, in Gloucestershire; these gentlemen accompanied their Report with the results of the experiments made at Manchester, and given more at length in this 2nd Part; adding in particular those with respect to the strength and best forms of iron beams and pillars.

166. Mr. Cubitt, likewise, feeling "impressed with the importance of further researches on the forms of cast iron beams, whether for the purpose of confirming or of extending the views hitherto taken," (Report, page 7,) caused eight experiments to be made on a moderately large scale. These experiments were published with the Report, and an abstract of them is in the following Table.

167. Tabular results of experiments on cast iron beams—each cast 16 feet long—laid, during the experiment, on supports 15 feet asunder, and intended to be of the same weight. The first six beams were uniform throughout, the seventh and eighth tapered towards the ends.

Table to Mr. Cubitt's Experiments.

	1	2	3	4	5	6	7	8	9	10	11	12	13	14	15	16	17
	Number.	Weight.	Size of bottom flange.	Size of top flange.	Thickness between flanges.	Total depth.	Area of section.	Distance of bearings.	Pressure applied to middle.	Deflexion.		Set.		Comparative stiffness.	Comparative power to resist impact.	Comparative strength reduced to equal area.	Comparative strength reduced to equal weight.
	1	cwt. qrs. lbs. 6 2 14	5·02 by 1·59	2·58 by ·86	·86 at top, 1·22 at bottom.	7·15	15·39	ft. in. 15 0	tons. 1 3 5 6 7	·265 ·73 1·285 1·54 1·80	Broke	·11 ·075 ·15 ·185		100	100	100	100
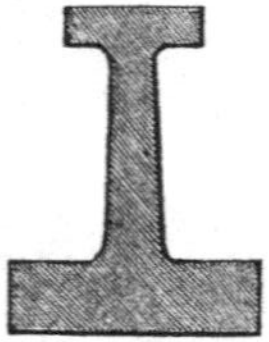	2	6 3 1	5·1 by 1·59	2·6 by ·86	·92 at top, 1·28 at bottom.	7·17	15·84	15 0	1 3 5 7 $7\frac{1}{8}$	·265 ·72 1·215 1·76 1·79	Broke	·015 ·058 ·121 ·243		104·6	101·2	98·9	99·7
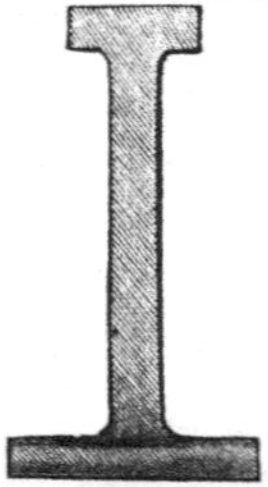	3	6 3 4	5·05 by 1·04	2·58 by ·87	·93	10·75	16·02	15 0	1 3 5 7 9 11 $11\frac{1}{2}$	·083 ·28 ·455 ·645 ·86 1·06 1·11	Broke	·02 ·04 ·058 ·08 ·105 ·14		265·8	101·3	157·8	160·4

			by 1·05	by ·88					3 5 7 9 11 12	·275 ·47 ·65 ·845 1·04 1·134	 Broke	·014 ·031 ·0545 ·085 ·124					
	5	6 0 24	5·08 by ·88	2·58 by ·65	·73	12·75	14·64	15 0	1 3 5 7 9 10 $10\frac{1}{4}$	·05 ·18 ·305 ·465 ·60 ·685 ·702	 Broke.	Set. ·137 ·0267 ·0525 ·0737 ·0852	 Bottom flange unsound.	420·8	57·0	153·9	156·1
	6	7 2 22	5·13 by 1·09	2·64 by ·94	·95	12·8	18·60	15 0	1 3 5 7 9 11 13 15 $15\frac{3}{4}$	·062 ·16 ·265 ·395 ·51 ·645 ·77 ·9 ·945	 Broke	·004 ·012 ·025 ·04 ·055 ·07 ·09 ·13		469·7	118·1	186·2	193·7

TABLE—*continued.*

1	2	3	4	5	6	7	8	9	10	11	12	13	14	15	16	17
Number.	Weight.	Size of bottom flange.	Size of top flange.	Thickness between flanges.	Total depth.	Area of section.	Distance of bearings.	Pressure applied to middle.	Deflexion.		Set.		Comparative stiffness.	Comparative power to resist impact.	Comparative strength reduced to equal area.	Comparative strength reduced to equal weight.
	cwt. qrs. lbs.						ft. in.	tons.								
7	6 2 18	6·5 by 1·	None.	·875 at top, ·95 at bottom.	14·	18·51	15 0	2 6 10 12 12⅜	·15 ·41 ·94 ·97	Broke.	·017 ·055 ·08 ·11	Bottom flange unsound.	360·7	95·2	147·	177·7
8	5 3 24	5·9 by ·84	2·72 by ·83	·68	17·25	18·11	15 0	2 6 10 14 16	·075 ·26 ·46 ·66 ·76	Broke.	·075	The set was not taken in this experiment, but it was discernible at ·1 ton. Bottom flange unsound.	631·2	96·5	194·2	253·9

	9	. .	5·05 by 1·59	2·6 by ·9	·87 at top, 1·26 at bottom.	7·15	15·63	7	6	2	·05	
										6	·185	
										10	·31	
										14	·47	
										$15\frac{5}{8}$	·52	Broke
	10	. .	5·06 by 1·02	2·6 by ·88	·92	10·75	15·89	7	6	4	·05	
										8	·105	
										12	·155	
										16	·21	
										20	·261	
										22	·29	
										$23\frac{7}{8}$	·318	Broke, nearly but not quite sound.
	11	. .	. .	. .	. .	. .	. .	7	6	4	·025	
										12	·085	
										20	·145	
										28	·22	
										31	·244	This piece (part of No. 6) was not broken, therefore no area is given.

168. Mr. Cubitt states that his object in making these experiments "was to show the great difference of strength of cast iron that can be got by taking certain forms." (Appendix to Report, page 39.) This object he has realized in a certain manner, as will be seen from the last two columns in the preceding Table, in which the beams, though of equal weight and section, are successively made to increase rapidly in strength above the top ones, which are the weakest.

169. Now as these weakest beams are not very different in form of section from, though somewhat weaker than, those which I have arrived at from a long course of inductive experiments, and considered as nearly the strongest in cast iron (arts. 135 and 136); and as the circumstance may be considered as bearing deeply upon the character of my published results with respect to beams, it will be incumbent on me to analyze the effort of Her Majesty's Commissioner with more freedom than I would otherwise willingly have done.

170. Speaking plainly, then, it appears to me that Mr. Cubitt, by increasing the strength through increasing the depth, the area being the same, has shown nothing more than would have been predicted from the slightest knowledge of theory; and that several of his beams, instead of showing greater strength, exhibit only weakness and inferiority of form.

171. To give a simple illustration of this statement, we will suppose a number of rectangular

beams to be formed, of the same length and area of section, but of different breadths and depths. Then the strength of each being as $b\,d^2$, varies as d, since $b\,d$, the area of the section, is constant.

172. We will select from Mr. Cubitt's Table the results of the different experiments, and attach to them a column containing the results which would have been derived from rectangular beams, of the same length and area, and varying in depth as Mr. Cubitt's did. It must, however, be understood that the rectangular section, which is comparatively a weak one, is introduced only for illustration, as its strength for different depths is easily computed; and it may be presumed that the strength of other sections, not so easily calculated, would increase, by augmenting the depth, in some such ratio as that does.

Number of Experi-ment.	Total depth of beam in middle.	Comparative strength reduced to equal areas.	Comparative strength of rectangular beam of the same depth as Mr. Cubitt's, and of constant area and length.	Remarks.
1	7·15	100	100	The strength of the first rectangular beam is assumed as 100, same as the first of Mr. Cubitt's.
2	7·17	98·9	100·27	
3	10·75	157·8	150·3	
4	10·75	152·4	150·3	
5	12·75	153·9*	178·3	* Bottom flange unsound.
6	12·8	186·2	179·0	
7	14·0	147*	195·8	* The bottom flange of both beams seems to have been slightly defective.
8	17·25	194·2*	241·2	

173. If we compare the results in the third and fourth columns, showing the comparative strengths of Mr. Cubitt's beams and of rectangular ones of the same depth, we shall see that the increased strength of Mr. Cubitt's beams, in the lower part of the Table, above the strength of that at the top, with which they are compared, is derived wholly from the *depth;* and as his latter beams give generally much lower results than are obtained from the rectangular section, we may infer that they are of inferior forms to that with which they are compared. The defect in the bottom flange of three out of four of them (an uncommon occurrence in properly cast beams) renders it probable that, if sound, they would have borne a little more than they did; but affords no probability that their increase of strength would have been equal to that of the rectangular section; which no doubt would have been the case, or nearly so, if the form in the first experiment had been used.

174. A further confirmation of the conclusion here arrived at is derived from the fact, that the strength of beams of the best form was found from my experiments to be, *cæteris paribus*, nearly as the depth; and the material in the section was but little increased with a large addition to the depth.

175. From the experiments and reasoning above Mr. Cubitt has drawn the conclusion of "*our knowledge of the best forms and arrangements of cast iron beams not being based upon principles the correctness*

of which cannot be questioned, (Report, page 10,) *and they are offered in confirmation or extension.*"

176. It would appear that Mr. Cubitt had mistaken the object of my experiments "on the strength and best forms of cast iron beams." It was virtually to seek for the form into which a given quantity of iron could be cast, so as to bear the greatest weight, the length and the depth of the beam being constant. Mr. Cubitt's experiments seem to have been intended to show that a given quantity of iron cast into beams—all of the same length—the section being of various forms (as the ⊥ section which I had represented as comparatively weak), would be made to bear more than others which I had represented as approaching to the strongest; this being effected merely by increasing the depth.

177. A little more attention to theoretical considerations might equally well have shown that increasing the depth—a privilege I did not allow myself when seeking for the best form of beam—had a great influence on the strength; and this might perhaps have prevented Mr. Cubitt offering to the public, under Her Majesty's sanction, additional examples, on a large scale, of weak beams.

178. In the tabular extract of Mr. Cubitt's experiments I have given the sets or defects of elasticity as obtained by that gentleman; but the length of his beams was not a sufficient number of times their depth for the results of the early sets, however carefully taken, to be any thing but an approximation.

179. As Mr. Cubitt's experimental results with respect to the strength of beams seem to be in accordance with my own, and might generally be computed from them, whatever opinion he may have formed to the contrary, I see no reason to doubt that the best form of beam is obtained from the reasonings and experiments previously given (arts. 108 to 144); and according to which many thousands of tons have been cast. I am preparing to repeat the leading experiments in my former effort on beams, on a very large scale, and to extend them considerably, through the liberality of an Iron Company.

COMPARATIVE STRENGTH OF HOT AND COLD BLAST IRON.

180. Having, in conjunction with Mr. Fairbairn, been requested, by the British Association for the Advancement of Science, to ascertain by experiment the comparative strengths of irons made by a heated and a cold blast, I will give here the results from my 'Report on the Tensile, Crushing, and Transverse Strengths of several kinds of Iron,' (Brit. Assoc. vol. vi. 1838,) attaching to them, in conclusion, the results from Mr. Fairbairn's experiments, which were on the latter kind of strain.

181. As the modes in which the different kinds of experiment were made, and many of the results obtained, are given in the earlier pages of this work, it will not be necessary here to enter into detail upon that subject. I shall, therefore, content myself

with stating, that the experiments were made with great care; and in devising the apparatus, the utmost attention was paid to theoretical requirements.

182. Taking only the means from all the experiments, in the Report above mentioned, and attaching to each result a number, in a parenthesis, indicative of the number of experiments from which it has been derived, we have as follows:

Carron Iron, No. 2, (Scotch.)

	Cold blast.	Hot blast.	Ratio representing cold blast by 1000.	
Tensile strength in lbs. per square inch	16683 (2)	13505 (3)	1000 : 809	
Compressive (crushing) strength in lbs. per square inch; from specimens cut out of castings previously torn asunder . .	106375 (3)	108540 (2)	1000 : 1020	Mean 997
Crushing strength obtained from prisms of various forms . .	100631 (9)	100738 (5)	1000 : 1001	
Do. from cylinders	125403 (13)	121685 (13)	1000 : 970	
Transverse strength from all the experiments	.... (11)	.... (13)	1000 : 991	
Computed power to resist impact .	.... (9)	.... (9)	1000 : 1005	
Transverse strength of bars, 1 inch square, and 4 feet 6 inches between the supports, in lbs. .	476 (3)	463 (3)	1000 : 973	
Ultimate deflexion of do. in inches	1·313 (3)	1·337 (3)	1000 : 1018	
Modulus of elasticity in lbs. per square inch (Part I. art. 256)	17270500 (2)	16085000 (2)	1000 : 931	
Specific gravity	7066	7046 (5)	1000 : 997	

Devon Iron, No. 3, (Scotch.)

Tensile strength per square inch .	....	21907 (1)	....
Compressive strength do. . . .	....	145435 (4)	....
Transverse do. from the experiments generally	.... (5)	.... (5)	1000 : 1417
Power to resist impact	.... (4)	.... (4)	1000 : 2786
Transverse strength of bars, 1 inch square, and 4 feet 6 inches between the supports . . .	448 (2)	537 (2)	1000 : 1199
Ultimate deflexion, do.	·79 (2)	1·09 (2)	1000 : 1380
Modulus of elasticity, do. . . .	22907700 (2)	22473650 (2)	1000 : 981
Specific gravity	7295 (4)	7229 (2)	1000 : 991

Buffery Iron, No. 1, (*English.*)

	Cold blast.	Hot blast.	Ratio representing cold blast by 1000.
Tensile strength per square inch .	17466 (1)	13434 (1)	1000 : 769
Compressive strength do. . . .	93366 (4)	86397 (4)	1000 : 925
Transverse strength	.... (5)	.... (5)	1000 : 931
Power to resist impact	.... (2)	.... (2)	1000 : 962
Transverse strength of bars, 1 inch square, and 4 feet 6 inches between the supports	463 (3)	436 (3)	1000 : 942
Ultimate deflexion, do.	1·55 (3)	1·64 (3)	1000 : 1058
Modulus of elasticity, do. . . .	15381200 (2)	13730500 (2)	1000 : 893
Specific gravity	7079	6998	1000 : 989

Coed-Talon Iron, No. 2, (*Welsh.*)

	Cold blast.	Hot blast.	Ratio representing cold blast by 1000.
Tensile strength per square inch .	18855 (2)	16676 (2)	1000 : 884
Compressive strength do. . . .	81770 (4)	82739 (4)	1000 : 1012
Specific gravity	6955 (4)	6968 (3)	1000 : 1002

Carron Iron, No. 3, (*Scotch.*)

Tensile strength per square inch .	14200 (2)	17755 (2)	1000 : 1250
Compressive strength do. . . .	115442 (4)	133440 (3)	1000 : 1156
Specific gravity	7135 (1)	7056 (1)	1000 : 989

183. Abstract of the transverse strengths, and powers to bear impact, as obtained from the experiments on the three irons first mentioned in the preceding Table. The bars, of whatever form, were usually cast 5 feet long, and laid upon supports 4 feet 6 inches asunder. Those of 1 inch square being

the bars from which the comparative powers to bear impact were computed, had their deflexions, from different weights, very carefully observed up to the time of fracture; and as the measured dimensions of the bar usually differed a small quantity from those of the model, the results, both as to strength and deflexion, were reduced by computation to what they would have been if the bar had been exactly 1 inch square. A reduction of this nature was made in the results of all the experiments, except otherwise mentioned. The comparative power to bear impact was obtained by multiplying the breaking weight of a bar, 1 inch square, by its ultimate deflexion, the length being always the same; a mode which is admissible, as appears from my experiments on the power of beams to bear impact, (British Association, 5th Report.)

Carron Iron, No. 2, (Scotch.)

Form and dimensions of section of casting.	Strength of irons. Cold blast iron.	Strength of irons. Hot blast iron.	Strength of irons. Ratio of strengths. The strength of cold blast iron being represented as 1000.	Power to bear impact. Cold blast iron.	Power to bear impact. Hot blast iron.	Power to bear impact. Ratio. The power of cold blast iron being represented as 1000.
	Results reduced.					
Rectangular bar, 1 inch square	492	469	1000 : 953·2	686	677·2	1000 : 987·1
Ditto, ditto	509	456	1000 : 895·8	711	649·3	1000 : 913·2
" "	429	465	1000 : 1083·9	493	532·0	1000 : 1079·1
" "	449	475	1000 : 1057·9	1481	1598·7	1000 : 1079·4
" "	457	429	1000 : 938·7	2601	2744·2	1000 : 1055·0
Bar, 3 inches deep and 1 inch broad . . .	3750	3843	1000 : 1024·8	141	154	1000 : 1092·2
" 5 inches deep and 1 inch broad . . .	10362	{10957, 9149} 10053	1000 : 970·1	3391	3087	1000 : 910·3
				530	452	1000 : 852·8
				359	458·6	1000 : 1277·4
		Mean	1000 : 989·1		Mean	1000 : 1005·1
	Results not reduced.					
Bar whose section is T	266	280	1000 : 1052·6			
Bar from the same model, but reversed	1050	980	1000 : 933·3			
Bar, section an isosceles triangle ▽	815	{672, 817} 742	1000 : 910·4			
Bar, a frustrum of ditto ▽	677	728	1000 : 1075·0			
		Mean	1000 : 992·8			
		General Mean	1000 : 990·9			

Devon No. 3 *Iron, (Scotch.)*

Form and dimensions of section of casting.	Strength of irons.			Power to bear impact.		
	Cold blast iron.	Hot blast iron.	Ratio of strengths. The strength of cold blast iron being represented as 1000.	Cold blast iron.	Hot blast iron.	Ratio. The power of cold blast iron being represented as 1000.
Bar, 1 inch square	448	504	1000 : 1125·0	353·9	589·2	1000 : 1664·8
„ 1 „	448	570	1000 : 1272·3	489·5	1761·7	1000 : 3598·9
„ 1½ inch deep by 1 inch broad . . .	890	1456	1000 : 1635·9	921·8	2747	1000 : 2980·0
				1702·3	4935	1000 : 2899·0
„ 3 „ by 1 „ . . .	3389	5183	1000 : 1529·3			
„ 5 „ by 1 „ . . .	10133	15422	1000 : 1521·9		Mean	1000 : 2785·6
		Mean	1000 : 1416·9			

Buffery No. 1 *Iron, (English, near Birmingham.)*

Form and dimensions of section of casting.	Cold blast iron.	Hot blast iron.	Ratio of strengths.	Cold blast iron.	Hot blast iron.	Ratio.
Bar, 1 inch square	491	464	1000 : 945·0	721·19	721·5	1000 : 1000·4
„ 1 „	437	437	1000 : 1000·0	2341·6	2163·2	1000 : 923·8
„ 1 „	462	409	1000 : 885·7			
„ 2 „	3057	2975	1000 : 973·1		Mean	1000 : 962·1
„ 2 „	3424	2903	1000 : 850·1			
		Mean	1000 : 930·7			

GENERAL SUMMARY OF TRANSVERSE STRENGTHS, AND COMPUTED POWERS TO RESIST IMPACT.

184. Selecting, from the irons above, the results of the experiments on the transverse strength, and power to resist impact, of the different bars broken, and adding to them the results of Mr. Fairbairn's experiments (Brit. Assoc. vol. vi.), we have as below.

Distance between supports 4 *feet* 6 *inches.*

Description of iron.	Strength of cold blast.	Strength of hot blast.	Ratio of strength, cold blast = 1000.	Ratio of powers to sustain impact, representing that of cold blast by 1000.
Carron, No. 2.	lbs.	lbs.		
Results from bars 1 inch square . . .	476 (3)	463 (3)	1000 : 973	
,, from all the experiments . . .	(11)	(13)	1000 : 991	1000 : 1005
Devon, No. 3.				
Results from bars 1 inch square . . .	448 (2)	537 (2)	1000 : 1199	
,, from all the experiments . . .	(5)	(5)	1000 : 1417	1000 : 2786
Buffery, No. 1.				
Results from bars 1 inch square . . .	463 (3)	436 (3)	1000 : 942	
,, from all the experiments . . .	(5)	(5)	1000 : 931	1000 : 962
Muirkirk, No. 1.				
Results from bars 1 inch square } Mr. Fairbairn's experiments.	454·2 (4)	418·9 (4)	1000 : 922	1000 : 823
Coed-Talon, No. 2, ditto	412·6 (5)	416·8 (4)	1000 : 1010	1000 : 1234
Coed-Talon, No. 3, ditto	553·2 (4)	513·1 (4)	1000 : 927	1000 : 925
Carron, No. 3, ditto	445·7 (5)	525·7 (5)	1000 : 1179	1000 : 1201
Elsicar, cold, and Milton, hot, No. 1, } ditto	451·5 (4)	369·4 (4)	1000 : 818	1000 : 875

185. These Tables contain the results of a very large number of experiments, made, with great care, upon English, Welsh, and Scotch iron, mostly supplied from the makers. They show that the irons marked No. 1, which are softer and richer than those of Nos. 2 and 3, are injured by the heated blast; since the hot blast irons of this description are less capable than the cold blast ones to resist fracture, whether the forces are tensile, compressive, transverse, or impulsive.

186. The irons marked No. 2, being harder than those of No. 1, have much less difference in their strength than the latter. In the Carron iron, No. 2, on which a great many experiments were made, the transverse strength, of the hot and cold blast specimens, was as 99 : 100, and their power of bearing impact equal. The Coed-Talon iron of this No. gave the transverse strength, of hot and cold blast, as 101 : 100, and their power to bear impact as 123 : 100. In both the Carron and the Coed-Talon irons, the hot blast castings were of equal strength to the cold blast ones, to resist crushing; but, in both, the strength of the hot blast was less than that of the cold blast, to resist tension, in a ratio of 8 or 9 to 10.

187. The No. 3 irons seem, both in aspect and strength, to be generally benefited by the heated blast. In the Carron iron, No. 3, the hot blast was superior to the cold, in the power of resisting tension, compression, transverse strain, and impact, in

a ratio approaching, in each case, that of 12 to 10 The Coed-Talon iron of this No. had, however, its hot blast kind weaker than the cold, to resist transverse strain and impact, in the ratio of about 93 to 100.

The iron, No. 3, from the Devon works in Scotland, was weak and irregular in the cold blast castings; but the hot blast iron from the same works was among the strongest I have tried. In this the ratio of the powers, of hot and cold blast iron, to bear pressure, was as 14 : 10; and to bear impact, as 28 : 10 nearly.

188. In these experiments, the hot blast irons usually differed from the cold blast, only so far as a different mode of manufacture—the introduction of a heated blast with coal, instead of a cold blast with coke—would produce. The difficulty we experienced in obtaining from the makers irons of both kinds made from the same materials, rendered it necessary to make the experiments on a smaller number of irons than would otherwise have been tried; but, from the evidence adduced, we may perhaps conclude that the introduction of a heated blast, into the manufacture of cast iron, has injured the softer irons, whilst it has frequently mollified and improved those of a harder nature; and considering the small deterioration which the irons of the quality No. 2 have sustained, and the apparent benefit to those of No. 3, together with the great saving effected by the heated blast, there seems good reason

for the process becoming as general as it has done. It is, however, to be feared that the facilities which the heated blast gives, of adulterating cast iron by mixture, have introduced into use a species of metal very inferior to that used in this comparison, or that from which the formulæ and leading results of this work have been obtained.

THEORETICAL INQUIRIES WITH REGARD TO THE STRENGTH OF BEAMS.

189. In the course of our remarks on the transverse strength of cast iron, as deduced from experiment, it was shown that the formulæ given by Tredgold, in Part I. of this work, were usually inapplicable to the computation of the strength of that metal to resist fracture. That very ingenious writer—following in the track of Dr. Young, and himself followed by numerous others—considered bodies, when not overstrained, to be perfectly elastic; and to resist extension and compression with equal energy. But theories deduced from these suppositions, however elegant, and nearly correct for small displacements of the fibres or particles, give the breaking strength of cast iron, in some cases, not half what it has been shown to bear by experiment (arts. 150 and 151). A square bar, instead of having its neutral line in the centre—one-half being extended and the other compressed, according to the suppositions above—requires to be considered as totally

incompressible, the neutral line being close to the side, or even beyond it,—a matter practically impossible. This defect in the received theories has been shown to arise from the neglect by writers of an element which appears always to be conjoined with elasticity, diminishing its power. This element —ductility, producing defective elasticity—will be attempted to be introduced into the following investigation; but as the formulæ are generally complex, and require additional experiments to supply their constant co-efficients, the reader may perhaps take for practical use the approximate ones previously given in this Second Part.

190. To find the position of the neutral line and the strength of a cast iron beam, supported at the ends, and loaded in the middle; the form of a section of the beam in the middle being that of the figure A B D E, where B C, H E, represent sections of the top and bottom ribs, F G that of the vertical one connecting them, and N O passes through the neutral line.

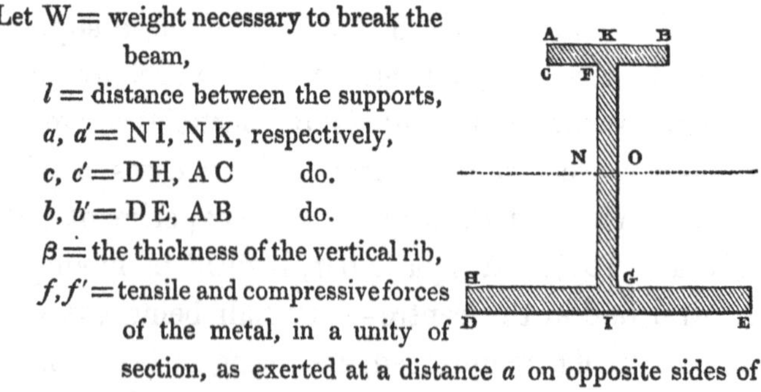

Let W = weight necessary to break the beam,
l = distance between the supports,
a, a' = N I, N K, respectively,
c, c' = D H, A C do.
b, b' = D E, A B do.
β = the thickness of the vertical rib,
f, f' = tensile and compressive forces of the metal, in a unity of section, as exerted at a distance a on opposite sides of the neutral line,

$\phi(x)$, $\phi'(x')$ = quantities respectively proportional to the forces of extension and compression of a particle, at a distance x from the neutral line,

n, n' = constant quantities dependent on the destruction of the elasticity of the material, by tensile and compressive forces.

191. The bottom rib will be in a state of tension, and the top one in a state of compression; and the parts of the section generally will be extended or compressed according to their distance from the line N O.

1st. To find the position of the neutral line.

192. Since $f.\frac{\phi(x)}{\phi(a)}$ = the force of the extended fibres or particles in a unity of section, at a distance x from the neutral line; therefore, multiplying this quantity by $b\,dx$, or by $\beta\,dx$, we have the force of the particles in an area of the section of which the breadth is b or β, and depth dx. Now the forces of tension, or those exerted by the particles below the line N O, may be expressed in two functions, the first representing the forces of the particles in the section N G, and the other those in the section H E.

$$\therefore f.\frac{\beta}{\phi(a)}\int_0^{a-c}\phi(x)\,dx + f.\frac{b}{\phi(a)}.\int_{a-c}^{a}\phi(x)\,dx = \text{S} = \text{sum of the forces of tension} \quad . \quad . \quad . \quad . \quad . \quad . \quad (1)$$

Proceeding in the same manner with respect to the compressed particles, we have

$$f' \cdot \frac{\beta}{\phi'(a)} \cdot \int_0^{a'-c'} \phi'(x')\,dx' + f' \cdot \frac{b'}{\phi'(a)} \cdot \int_{a'-c'}^{a'} \phi'(x')\,dx' = \mathrm{S}' = \text{sum of the forces of compression} \quad . \quad . \quad . \quad . \quad (2)$$

The weight acts in a direction parallel to the section of fracture, and therefore the sum of the forces of extension and compression, being the only horizontal forces, are equal to each other.

$$\therefore \mathrm{S} = \mathrm{S}' \quad . \quad . \quad . \quad . \quad . \quad . \quad . \quad . \quad (3)$$

193. It appears from my recent experiments that no rigid body is perfectly elastic; and it has been shown (art. 99 to 106), in extracts from communications which I made to the York and late Cambridge meetings of the British Association, that, in the flexure of bars of cast iron and stone, the defect of elasticity was nearly as the square of the weight applied, or of the deflexion, though the defect from the smaller deflexions seemed to increase in a somewhat lower ratio. Other experiments on the defect of elasticity, as exhibited in the flexure of bars of cast iron, wrought iron, steel, timber, and stone,—and on the defect of elasticity in the longitudinal variations of bodies—will appear in a future volume of the British Association. As the latter experiments, those on longitudinal variation, are not at present completed, I shall assume $\phi(x) = x - \frac{x^v}{n\,a}$, and $\phi'(x') = x' - \frac{x'^{v_\prime}}{n'\,a}$; where v, v', n, n' are supposed to be constant, and v, v' to represent the powers of x, x', to which the defects of elasticity of extension and compression are proportional.

Substituting in equations (1) and (2), the values of $\phi(x)$ and $\phi'(x')$, as above, we have

$$f.\frac{n\beta}{na-a^{v-1}}.\int_{0}^{a-c}\left(x-\frac{x^v}{na}\right)dx+f.\frac{nb}{na-a^{v-1}}.\int_{a-c}^{a}\left(x-\frac{x^v}{na}\right)dx=\mathrm{S} \quad . \quad (4)$$

$$f'.\frac{n'\beta}{n'a-a^{v'-1}}.\int_{0}^{a'-c'}\left(x'-\frac{x'^{v'}}{n'a}\right)dx'+f'.\frac{n'b'}{n'a-a^{v'-1}}.\int_{a'-c'}^{a'}\left(x'-\frac{x'^{v'}}{n'a}\right)dx'=\mathrm{S}' \quad . \quad . \quad (5)$$

Performing the integrations of equations (4) and (5) gives

$$f.\frac{n\beta}{na-a^{v-1}}.\int_{0}^{a-c}\left(x-\frac{x^v}{na}\right)dx=\frac{fn\beta}{na-a^{v-1}}\left(\frac{(a-c)^2}{2}-\frac{(a-c)^{v+1}}{(v+1)\,na}\right),$$

$$f.\frac{nb}{na-a^{v-1}}.\int_{a-c}^{a}\left(x-\frac{x^v}{na}\right)dx=\frac{fnb}{na-a^{v-1}}\left\{\left(\frac{a^2}{2}-\frac{a^{v+1}}{(v+1)\,na}\right)-\left(\frac{(a-c)^2}{2}-\frac{(a-c)^{v+1}}{(v+1)\,na}\right)\right\}$$

$$f'.\frac{n'\beta}{n'a-a^{v'-1}}.\int_{0}^{a'-c'}\left(x'-\frac{x'^{v'}}{n'a}\right)dx'=\frac{f'n'\beta}{n'a-a^{v'-1}}\left(\frac{(a'-c')^2}{2}-\frac{(a'-c')^{v'+1}}{(v'+1)\,n'a}\right)$$

$$f'.\frac{n'b'}{n'a-a^{v'-1}}.\int_{a'-c'}^{a'}\left(x'-\frac{x'^{v'}}{n'a}\right)dx'=\frac{f'n'b'}{n'a-a^{v'-1}}\left\{\left(\frac{a'^2}{2}-\frac{a'^{v'+1}}{(v'+1)\,n'a}\right)-\left(\frac{(a'-c')^2}{2}-\frac{(a'-c')^{v'+1}}{(v'+1)\,n'a}\right)\right\}.$$

Collecting together the integrals of equation (4), we have

$$\mathrm{S}=\frac{fn}{na-a^{v-1}}\left\{\frac{ba^2}{2}-\frac{ba^{v+1}}{(v+1)\,na}-\frac{b\,(a-c)^2}{2}+\frac{\beta(a-c)^2}{2}+\frac{b(a-c)^{v+1}}{(v+1)\,na}-\frac{\beta\,(a-c)^{v+1}}{(v+1)\,na}\right\}$$

$$=\frac{fn}{na-a^{v-1}}\left\{ba^2\left(\frac{1}{2}-\frac{a^{v-1}}{(v+1)\,na}\right)-(b-\beta)\left(\frac{(a-c)^2}{2}-\frac{(a-c)^{v+1}}{(v+1)\,na}\right)\right\}.$$

In like manner,

$$S' = \frac{f' n'}{n'a - a^{v'-1}} \left\{ b' a'^2 \left(\frac{1}{2} - \frac{a'^{v'-1}}{(v'+1)\, n'a} \right) - (b'-\beta) \left(\frac{(a'-c')^2}{2} - \frac{(a'-c')^{v'+1}}{(v'+1)\, n'a} \right) \right\}.$$

Equating the values of S and S′ gives, for the equation of the neutral line,

$$\frac{f}{n - a^{v-2}} \left\{ ba^2 \left(\frac{n}{2} - \frac{a^{v-1}}{(v+1)\, a} \right) - (b-\beta) \left(\frac{n\,(a-c)^2}{2} - \frac{(a-c)^{v+1}}{(v+1)\, a} \right) \right\} = \frac{f'}{n' - a^{v'-2}}$$
$$\left\{ b'a'^2 \left(\frac{n'}{2} - \frac{a'^{v'-1}}{(v'+1)\, a} \right) - (b' - \beta) \left(\frac{n'\,(a'-c')^2}{2} - \frac{(a'-c')^{v'+1}}{(v'+1)\, a} \right) \right\} \quad (6)$$

where $a' = D - a$, D being the depth of the beam.

Cor. 1st. If $v = 2$, and $v' = 2$, as would appear to be nearly the case from the experiments on the transverse flexure of cast iron bars, mentioned above, the equation of the neutral line would be

$$\frac{f}{n-1} \left\{ ba^2\,(3\,n-2) - (b-\beta) \left(3\,n\,(a-c)^2 - \frac{2\,(a-c)^3}{a} \right) \right\} = \frac{f'}{n'-1}$$
$$\left\{ b'\,a'^2 \left(3\,n' - \frac{2\,a'}{a} \right) - (b'-\beta) \left(3\,n'\,(a'-c')^2 - \frac{2\,(a'-c')^3}{a} \right) \right\} \quad (7)$$

If the beam is of the ⊥ form, having no top rib, then $c' = 0$, $b' = 0$, and equation (7) becomes

$$\frac{f}{n-1} \left\{ ba^2\,(3\,n-2) - (b-\beta) \left(3\,n\,(a-c)^2 - \frac{2\,(a-c)^3}{a} \right) \right\} = \frac{f'\,\beta a'^2}{n'-1}$$
$$\left(3\,n' - 2\,\frac{a'}{a} \right) \quad (8)$$

If $b = b' = \beta$, then the section of the beam is rectangular, as a joist; and the equation of the neutral line (7) becomes

$$\frac{fa^2}{n-1}\,(3n-2) = \frac{f'a'^2}{n'-1} \left(3n' - 2\,\frac{a'}{a} \right) \quad . \quad . \quad . \quad . \quad (9)$$

Cor. 2nd. If $v=1$, and $v'=1$, or the defect of elasticity be as the extension and compression, then equation (6) will become

$$\frac{f}{2\left(n-\frac{1}{a}\right)}\left\{ba^2\left(n-\frac{1}{a}\right)-(b-\beta)(a-c)^2\left(n-\frac{1}{a}\right)\right\}=\frac{f'}{2\left(n'-\frac{1}{a}\right)}$$

$$\left\{b'a'^2\left(n'-\frac{1}{a}\right)-(b'-\beta)(a'-c')^2\left(n'-\frac{1}{a}\right)\right\}\therefore\frac{f}{2}\left\{ba^2-(b-\beta)\right.$$

$$\left.(a-c)^2\right\}=\frac{f'}{2}\left\{b'a'^2-(b'-\beta)(a'-c')^2\right\} \quad . \; . \; . \; . \; . \; . \; . \; . \quad (10)$$

But this equation becomes

$$f\left\{ba\times\frac{a}{2}-(b-\beta)(a-c)\times\frac{a-c}{2}\right\}=f'\left\{b'a'\times\frac{a'}{2}-(b'-\beta)(a'-c')\times\frac{a'-c'}{2}\right\},$$

where $b\,a$ = area of the whole section of tension considered as a rectangular surface of which the breadth is b and depth a; $\frac{a}{2}$ = distance of centre of gravity of that section from the neutral line; $(b-\beta)$ $(a-c)$ = area of part necessary to complete the rectangular surface above; $\frac{a-c}{2}$ = distance of centre of gravity of this defective part. In like manner $b'\,a'$ = area of whole section of compression considered as a rectangular surface of breadth b' and depth a'; $\frac{a'}{2}$ = distance of its centre of gravity; $(b'-\beta)$ $(a'-c')$ = area of part wanting; and $\frac{a'-c'}{2}$ = the distance of its centre of gravity.

Whence it appears that if $f=f'$, the neutral line will be in the centre of gravity of the section.

Cor. 3rd. If the elasticity be considered as perfect, then n, n' will be infinitely great, and taking as usual $f=f'$, equation (6) will become

$$ba^2-(b-\beta)(a-c)^2=b'a'^2-(b'-\beta)(a'-c')^2 \quad . \quad . \quad (11)$$

The curious circumstance of this equation being in agreement with equation (10), is rendered obvious by other reasoning. For, in Cor. 2, we have

$$\phi(x)=x-\frac{x}{na}=\left(1-\frac{1}{na}\right)x, \text{ and } \phi'(x')=x'-\frac{x'}{n'a}=\left(1-\frac{1}{n'a}\right)x',$$

the forces being as the extensions and compressions.

If, on the supposition of perfect elasticity, $b'=0$, and $c'=0$, the beam being of the ⊥ form of section, having no top rib, the last equation will become

$$ba^2-(b-\beta)(a-c)^2=\beta a'^2 \quad . \quad . \quad . \quad (12)$$

If the beam be rectangular and perfectly elastic, then $b=b'=\beta$, and equation (11) becomes

$$ba^2=b'a'^2$$
$$\therefore a^2=a'^2,$$

or the neutral line is in the middle.

2nd. To find the strength of the beam, the values of a, a', and consequently the position of the neutral line, having been previously determined, from one of the preceding equations, or by other means.

194. Since, by equation (4) of the preceding article, the sum of the forces of extension is

$$\frac{fn\beta}{na-a^{v-1}}\cdot\int_0^{a-c}\left(x-\frac{x^v}{na}\right)dx+\frac{fnb}{na-a^{v-1}}\cdot\int_{a-c}^{a}\left(x-\frac{x^v}{na}\right)dx.$$

And as the moment of each of these forces, with respect to the neutral line, is equal to the product of the force by its distance x from that line, we have for the sum of the moments of the forces of tension,

$$\frac{fn\beta}{na-a^{v-1}}\cdot\int_{o}^{a-c}\left(x^2-\frac{x^{v+1}}{na}\right)dx+\frac{fnb}{na-a^{v-1}}\cdot\int_{a-c}^{a}\left(x^2-\frac{x^{v+1}}{na}\right)dx=S_{,}\ \ldots\ (13)$$

$$\therefore\frac{fn\beta}{na-a^{v-1}}\cdot\int_{o}^{a-c}\left(x^2-\frac{x^{v+1}}{na}\right)dx=\frac{fn\beta}{na-a^{v-1}}\cdot\left\{\frac{(a-c)^3}{3}-\frac{(a-c)^{v+2}}{(v+2)\,na}\right\},$$

and

$$\frac{fnb}{na-a^{v-1}}\cdot\int_{a-c}^{a}\left(x^2-\frac{x^{v+1}}{na}\right)dx=\frac{fnb}{na-a^{v-1}}\cdot\left\{\frac{a^3}{3}-\frac{a^{v+2}}{(v+2)\,na}-\frac{(a-c)^3}{3}+\frac{(a-c)^{v+2}}{(v+2)\,na}\right\}$$

Whence we have, for the sum of the moments of the forces of the extended particles,

$$S_{,}=\frac{f}{3a\,(n-a^{v-2})}\cdot\left\{ba^3\left(n-\frac{3\,a^{v-2}}{v+2}\right)-(b-\beta)\left(n\,(a-c)^3-\frac{3(a-c)^{v+2}}{(v+2)\,a}\right)\right\}\ \ldots\ldots\ (14)$$

In like manner we obtain, for the sum of the moments of the forces of the compressed particles,

$$S_{,,}=\frac{f'}{3a(n'-a^{v'-2})}\cdot\left\{b'a'^3\left(n'-\frac{3a'^{v'-1}}{(v'+2)\,a}\right)-(b'-\beta)\left(n'\,(a'-c')^3-\frac{3\,(a'-c')^{v'+2}}{(v'+2)\,a}\right)\right\}\ \ldots\ldots\ (15)$$

But $S_{,}+S_{,,}$ the sum of the moments of the forces of extension and compression, must be equal to the product of half the weight laid on the middle of the beam by half the distance between the supports; for we may consider the beam as fixed firmly in the middle, and loaded at one end with half the weight laid on the middle.

$$\therefore S_{,} + S_{,,} = \tfrac{1}{2} W \times \tfrac{1}{2} l = \frac{W l}{4} \quad . \quad . \quad . \quad (16)$$

Cor. 1. If $c' = 0$, and $b' = 0$, the section of the beam being of the ⊥ form, having no top rib, we have from equation (15),

$$S_{,,} = \frac{f'\beta}{3a(n' - a^{v'-2})}\left(n'a'^{3} - \frac{3a'^{v'+2}}{(v'+2)a}\right),$$

and $S_{,}$ as in equation (14), for substitution in equation (16).

Cor. 2. If c, c', b, b' are each $= 0$, or the beam is rectangular, as a joist, we have from equation (14)

$$S_{,} = \frac{f\beta a^{2}}{3(n - a^{v-2})}\left(n - \frac{3a^{v-2}}{v+2}\right),$$

from equation (15)

$$S_{,,} = \frac{f'\beta a'^{3}}{3a(n' - a^{v'-2})}\left(n' - \frac{3a'^{v'-1}}{(v'+2)a}\right),$$

where

$$\frac{W l}{4} = S_{,} + S_{,,} \quad . \quad . \quad . \quad . \quad . \quad . \quad . \quad (17)$$

Cor. 3. If v, v' are each $= 2$, as would appear from the experiments (art. 106), then the values of $S_{,}$, $S_{,,}$, for the strength of a rectangular section in the last corollary, give

$$\frac{W l}{4} = \frac{f\beta a^{2}}{12(n-1)}(4n-3) + \frac{f'\beta a'^{3}}{12(n'-1)a}\left(4n' - \frac{3a'}{a}\right) \qquad (18)$$

where $\frac{a'}{a}$ is the ratio of the depth of the compressed section to that of the extended one.

On the supposition of this corollary, that $v = v' = 2$, the general formulæ in equations (14) (15) give

$$S_{,} = \frac{f}{12a(n-1)}\left\{ ba^3(4n-3) - (b-\beta)\left(4n(a-c)^3 - \frac{3(a-c)^4}{a}\right)\right\}.$$

$$S_{,,} = \frac{f'}{12a(n'-1)}\left\{ b'a'^3\left(4n' - \frac{3a'}{a}\right) - (b'-\beta)\left(4n'(a'-c')^3 - \frac{3(a'-c')^4}{a}\right)\right\}.$$

Cor. 4. If $v = v' = 1$, or the defect of elasticity is as the extension and compression, equations (14) and (15) give

$$S_{,} = \frac{f}{3a}\left\{ ba^3 - (b-\beta)(a-c)^3 \right\}, \text{ and}$$

$$S_{,,} = \frac{f'}{3a}\left\{ b'a'^3 - (b'-\beta)(a'-c')^3 \right\}$$

Cor. 5. If the beam be supposed to be perfectly elastic, then n and n' are both infinite, and we have from equations (14) and (15)

$$S_{,} = \frac{f}{3a}\left\{ ba^3 - (b-\beta)(a-c)^3 \right\},$$

$$S_{,,} = \frac{f'}{3a}\left\{ b'a'^3 - (b'-\beta)(a'-c')^3 \right\},$$

agreeing with the results of the last corollary, as previously remarked with respect to equations (10) and (11).

But as the body is elastic we will assume as usual $f = f'$,

$$\therefore S_{,} + S_{,,} = \frac{f}{3a}\left\{ ba^3 + b'a'^3 - (b-\beta)(a-c)^3 - (b'-\beta)(a'-c')^3 \right\} = \frac{Wl}{4} \quad .\ .\ (19)$$

If $b = b'$, and $c = c'$, or the top and bottom ribs

are equal, then, the neutral line being in the centre, $a = a'$; and the last equation gives

$$\frac{\mathrm{W}\,l}{4} = \frac{2f}{3a}\left\{ba^3 - (b-\beta)\,(a-c)^3\right\}, \quad . \quad . \quad . \quad (20)$$

a result in agreement with equation XIX., art. 116 of Tredgold.

If in this case $\beta = 0$, or the part between the ribs was so thin that it might be neglected,

$$\frac{\mathrm{W}\,l}{4} = \frac{2fb}{3a}\left\{a^3 - (a-c)^3\right\}, \quad . \quad . \quad . \quad . \quad (21)$$

agreeing with art. 117 of Tredgold.

If $\beta = b$, or the beam is rectangular, then equation (20) becomes

$$\frac{\mathrm{W}\,l}{4} = \frac{2f}{3a} \times ba^3 = \frac{2fba^2}{3} = \frac{fb\mathrm{D}^2}{6}, \quad . \quad . \quad . \quad (22)$$

where D is the whole depth $= 2\,a$, a result usually arrived at by a much simpler process (Tredgold, art. 110).

For other investigations on the subject of the neutral line, and the strength of beams variously fixed—the elasticity being supposed perfect—see Professor Moseley's 'Principles of Engineering and Architecture.'

RESISTANCE TO TORSION.

195. If a prismatic body, fixed firmly at one end, have a weight applied to twist it by means of a lever acting at the other, perpendicular to the length of the body, to find the resistance to twisting and to fracture.

196. The problem here proposed has been made the subject of an ingenious article by Tredgold in the 1st Part of this work; but as it has been subjected to more recent and profound theoretical investigation by Cauchy and others on the Continent, the formulæ given by Navier ('Application de Mécanique'), including those of Cauchy, will be inserted here, referring for their demonstrations to the work itself, or to M. Cauchy's 'Exercices de Mathématiques,' 4e année. Experimental results by Bevan, Rennie, Savart, and the Author, will likewise be given or noticed.

197. Let l = the length of the prism from the fixed end to the point of application of the lever used to twist it.

r = the radius of the prism, if round.

b, d = its breadth and thickness, if rectangular.

P = the weight acting by means of the lever to twist it.

R = the length of the lever.

θ = the angle of torsion, at the point of application, considered as very small.

G = a constant for each species of body, representing the specific resistance to flexure by torsion.

T = a constant weight expressing the resistance to torsion, with regard to a unit of surface, at the time of fracture.

We have then as in the following Table.

Form of section of prism.	Resistance to angular flexure by a force of torsion.	Resistance to fracture by a force of torsion.
Round . .	$G=PR.\frac{2l}{\pi r^4 \theta}.$	$T=PR.\frac{2}{\pi r^3}.$
Square. .	$G=PR.\frac{6l}{d^4 \theta}.$	$T=PR.\frac{6}{\sqrt{2}.d^3}$
Rectangular	$G=PR.\frac{3\ (b^2+d^2)\ l}{b^3\ d^3\ \theta}.$	$T=PR.\frac{3\sqrt{b^2+d^2}}{b^2.d^2}$

The formulæ for G are in art. 160, and those for T in art. 168, of the 'Application de Mécanique,' 1re partie; those for the rectangular prism being from Cauchy.

198. If E be the force necessary to elongate or shorten a prism, the transverse section of which is a superficial unity, as one square inch, by a quantity equal to the length of the prism, the elasticity being supposed perfect, and the force applied in the direction of the length; and if F be the force necessary to break such a prism; then E will be the modulus of elasticity of the material, and F the

modulus of its resistance to fracture; and the values of G and T above will be connected with those of E and F by the following relations.

$$G = \frac{2E}{5}, \quad T = \frac{4F}{5}.$$

See 'Application de Mécanique,' notes to articles 159 and 167.

The connexion is, however, but little in accordance with the results of experiment, with respect to the values of T and F, as might be expected from the elasticity being much injured previous to the time of fracture.

199. The angle θ being measured by the arc due to a radius equal unity, if the angle were expressed in degrees, and represented by Δ, we should have $\theta = \Delta \cdot \frac{\pi}{180} = \frac{\Delta}{57 \cdot 29578}$, where 57·29578, or $\frac{180}{\pi}$, is the number of degrees in the arc whose length is equal to radius.

200. To obtain the values of G and T in the preceding Table, with respect to any particular material, as cast iron, we must refer to experiment, and will next insert some results which were kindly sent by the author, Mr. Geo. Rennie, in 1842, as part of a more general inquiry; noticing afterwards other experiments previously mentioned, besides some of an earlier date, quoted by Tredgold.

201. Experiments, by Geo. Rennie, Esq., F.R.S., on the strength of three bars of cast iron to resist fracture by torsion. The bars were planed exactly

one inch square, and were firmly fixed, at one end, in a horizontal position, and broken by weights acting at the other by means of an arched lever, 3 feet in length, perpendicular to the bar, and exactly balanced by a counter weight. The bars were cast from the cupola, the first vertically, the other two horizontally. The former was broken with 191 lbs., and the two latter with 231 lbs. each.

202. To determine from the preceding experiments the value of T, the modulus of resistance to fracture by torsion, in the formula, $T = P R . \frac{6}{\sqrt{2} . d^3}$, for a square prism,—taking the weights in pounds, and the dimensions in inches,—we have R=36, $d=1$, and $T = P \times \frac{36 \times 6}{\sqrt{2}} = 152{\cdot}735 \times P$.

In the horizontal casting	$P=191, \therefore T=29172{\cdot}4$	}	Mean
In the vertical castings	$P=231, \therefore T=35281{\cdot}8$	}	32227 lbs.

If R were taken in feet and the rest in inches, and pounds, as before, the value of T would be 2685·6.

203. The experiments of Messrs. Bramah on the torsion of square bars of cast iron (Tredgold, art. 85) give for T, taking the dimensions in inches,

No. 3 . . .	T = 42020·4	
„ 4 . . .	T = 39474·0	Mean 37747 lbs.
„ 6 . . .	T = 36198·0	
„ 7 . . .	T = 33296·4	

204. Mr. Dunlop's experiments on the torsion of cylinders, varying, in diameter from 2 to $4\frac{1}{2}$ inches, and in length from $2\frac{3}{4}$ to 6 inches, the leverage being 14 feet 2 inches (Tredgold, art. 85), give, from the formula $T = P R . \frac{2}{\pi r^3}$, as follows:

In experiment No. 2, T = 27056·4		
,, 3, T = 29187·6		
,, 5, T = 29142·0		
,, 6, T = 29509·2	Mean 27534 lbs.	
,, 8, T = 27286·8		
,, 9, T = 26217·6		
,, 10, T = 24338·4		

Taking a mean from the three mean results last obtained gives, T = 32503 lbs.; the dimensions being in inches.

205. Putting this value for T, in the formula (art. 197), and transposing, we obtain the following value of P R.

In a cylinder . . . $PR = 51055 \cdot r^3$.

In a square prism . . $PR = 7661 \cdot d^3$.

In a rectangular prism, $PR = 10834 \cdot \frac{b^2 d^2}{\sqrt{b^2 + d^2}}$.

206. If R be taken in feet, as was supposed by Tredgold, art. 265, we shall have, for a square prism, $P = \frac{638 d^3}{R}$.

Hence his co-efficient, 150, being less than $\frac{1}{4}$th of that of fracture, may be regarded as perfectly safe for practical application.

207. Mr. Benjamin Bevan gave, in the 'Philosophical Transactions' for 1829, a memoir containing numerous experiments "on the modulus of torsion."

They were principally on timber, but contained a small Table of the modulus of torsion of metals. Mr. Bevan defined this modulus by the value of T in the following equation for a square prism, twisted as before described:

$$\frac{R^2 l P}{d^4 T} = \delta,$$

where δ is the deflexion, considered as very small, and the rest of the notation as before; the weights and the dimensions being in pounds and inches.

We have from above

$$\frac{R l P}{d^4 T} = \frac{\delta}{R};$$

and as $\frac{\delta}{R}$ is the deflexion at a unity of distance, and very small, it may be taken for the arc.

$$\therefore \frac{R l P}{d^4 T} = \theta, \text{ where } T = \frac{P R l}{d^4 \theta}$$

But from the Table (art. 197), $G = P R . \frac{6 l}{d^4 \theta}$ $= \frac{6 P R l}{d^4 \theta}, \therefore G = 6 T.$

208. Mr. Bevan finds the modulus T in cast iron, whose specific gravity is 7·163, to be as below.

$$\left.\begin{matrix} 940000 \\ 963000 \\ 952000 \end{matrix}\right\} \text{Mean } 951600 \text{ lbs.}$$

The moduli of wrought iron and steel were nearly equal to each other; and a mean from the results of eight experiments on iron and three on steel, gave, for T, 1779090 lbs.

209. Mr. Bevan found the modulus T to be $\frac{1}{16}$th of the modulus of elasticity in metallic substances.

But it was shown above that $G = 6 T$, $\therefore G = \frac{6}{16}$ of the modulus of elasticity; which differs from $\frac{2}{5}$, as computed by Cauchy (art. 198), only as 16 to 15.

210. Multiplying Mr. Bevan's mean values of T by six we obtain,

In cast iron, $G = 951600 \times 6 = 5709600$ lbs.

In wrought iron and steel, $G = 1779090 \times 6 = 10674540$ lbs.

211. Substitute in the formulæ (art. 197) the value of G, as obtained from the above experiments on cast iron, and transposing, we have as below.

For a cylinder . . . $PR = 8968620 \cdot \frac{r^4 \theta}{l}$.

For a square prism . . $PR = 951600 \cdot \frac{d^4 \theta}{l}$.

For a rectangular prism, $PR = 1903200 \cdot \frac{b^3 d^3 \theta}{(b^2 + d^2) l}$.

212. The experiments of Mr. Bevan seem to have been very carefully made, extra precaution being used both to prevent friction and to obtain correct measures; but a source of error may have arisen from computing, by a less accurate formula than Cauchy's, the results from those rectangular prisms which differed considerably from squares.

213. Some experiments of my own, upon the torsion of cylinders of wrought iron and steel, made, some years since, at the request of Mr. Babbage, but not yet published, showed that the angle of torsion was very nearly as the weight, as had previously been shown by Savart, (Annales de Chimie et de Physique, Aug. 1829.)

214. The object of M. Savart's able Memoir was to compare the theory of torsion, as given by Pois-

son and Cauchy, with the results of experiment; and though his experiments were made on brass, copper, steel, glass, oak, &c.—and included none on cast iron—it may not be amiss to give here the general laws which he deduces from them. They are as below.

1st. Whatever be the form of the transverse section of the rods (subjected to torsion), the arcs of torsion are directly proportional to the moment of the force and to the length.

2nd. When the sections of the rods are similar, whether circular, triangular, square, or rectangular, much elongated, the arcs of torsion are in the inverse ratio of the fourth power of the linear dimensions of the section.

3rd. When the sections are rectangles, and the rods possess an uniform elasticity in every direction, the arcs of torsion are in the inverse ratio of the product of the cubes of the transverse dimensions, divided by the sum of their squares; from whence it follows that, if the breadth is very great compared with the thickness, the arcs of torsion will be sensibly in the inverse ratio of the breadth and of the cube of the thickness.

These laws, M. Savart observes, are in exact agreement with those of Cauchy, both for cylindrical and rectangular sections; showing that his formulæ (art. 197) are constructed on principles which may be applied with safety. When the elasticity is not uniform, the laws are somewhat modified.

215. In concluding this notice of experiments on the strength and other properties of cast iron, which may, perhaps, on a future occasion be extended in various ways, I would refer the reader desirous of information on the effects of expansion and contraction, upon structures, by heat, to a Memoir 'on the Expansion of Arches' through the changes of ordinary temperature, by Mr. George Rennie. This Memoir contains experiments on the rise of the arches in the Southwark Bridge, which is of cast iron, having three rows of arches in length, containing in the whole about 5560 tons of iron. Mr. Rennie made experiments upon the rise of the arches in each row; from which it appears that the rise of an arch, whose span is 246 feet and versed sine 23 feet 1 inch, is about $\frac{1}{40}$th of an inch for each degree of Fahrenheit, making $1\frac{1}{4}$ inch for a difference of 50°. Mr. Rennie gives a Table of experiments of his own upon the expansion of iron and stone, with others from M. Destigny; and concludes that there is no more danger to the stability of iron bridges, from the effects of expansion and contraction, than to those of stone; for when the abutments are firmly fixed, the arches have no alternative but to rise or fall.

The effects of percussion and vibration upon bodies, particularly cast iron, have been much further inquired into since the time of Tredgold; and upon these subjects I beg to refer the reader to a Memoir of my own on the effects of "Impact upon Beams," in the 5th Report of the British Association, 1835. The object of this Memoir was to compare theory with experiment, deducing practical conclusions.

THE END.

PRINTED BY W. HUGHES, KING'S HEAD COURT, GOUGH SQUARE.

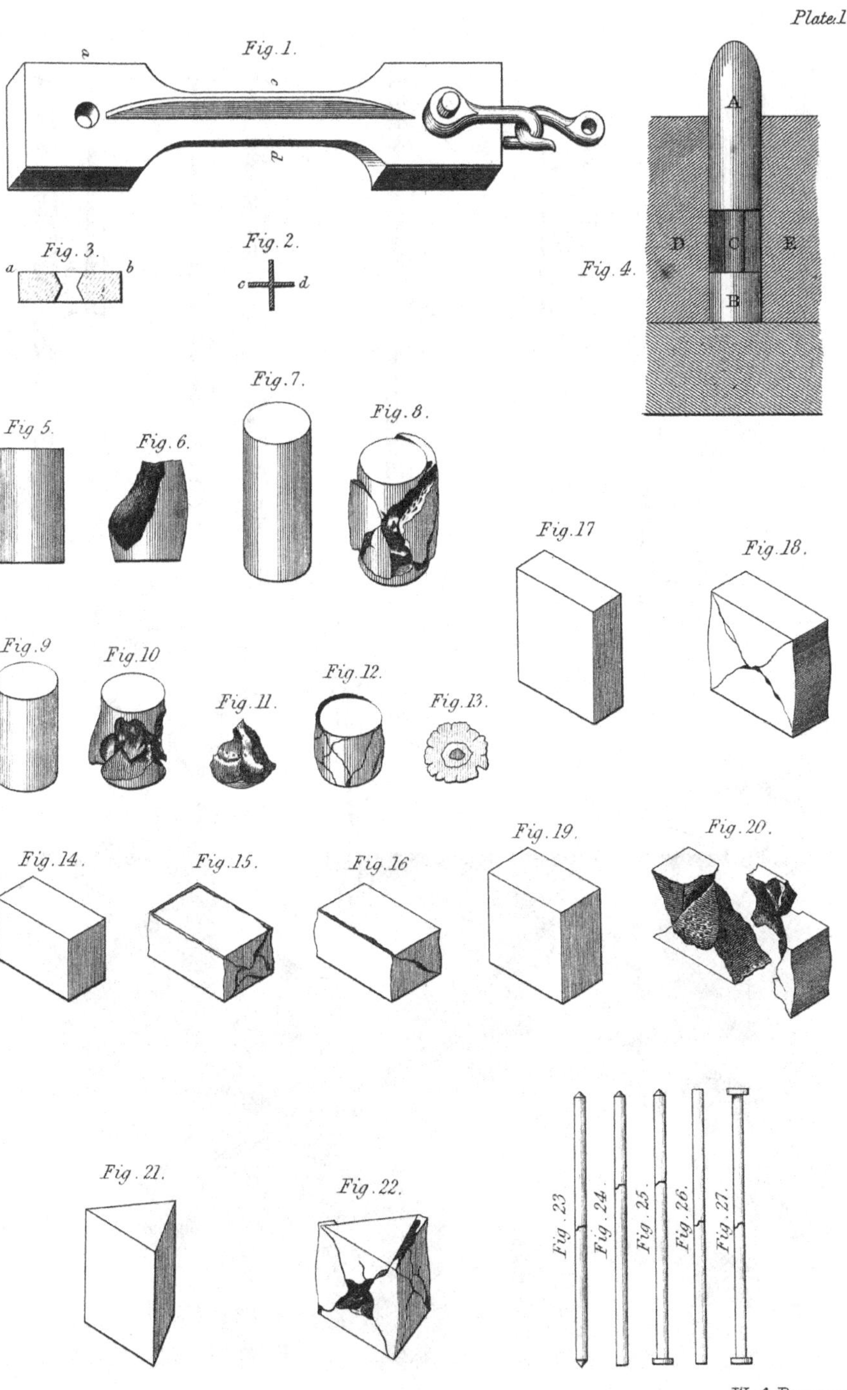

W. A. Beever. sc.

Published by J. Weale, at the Architectural Library, 59. High Holborn. 1846.

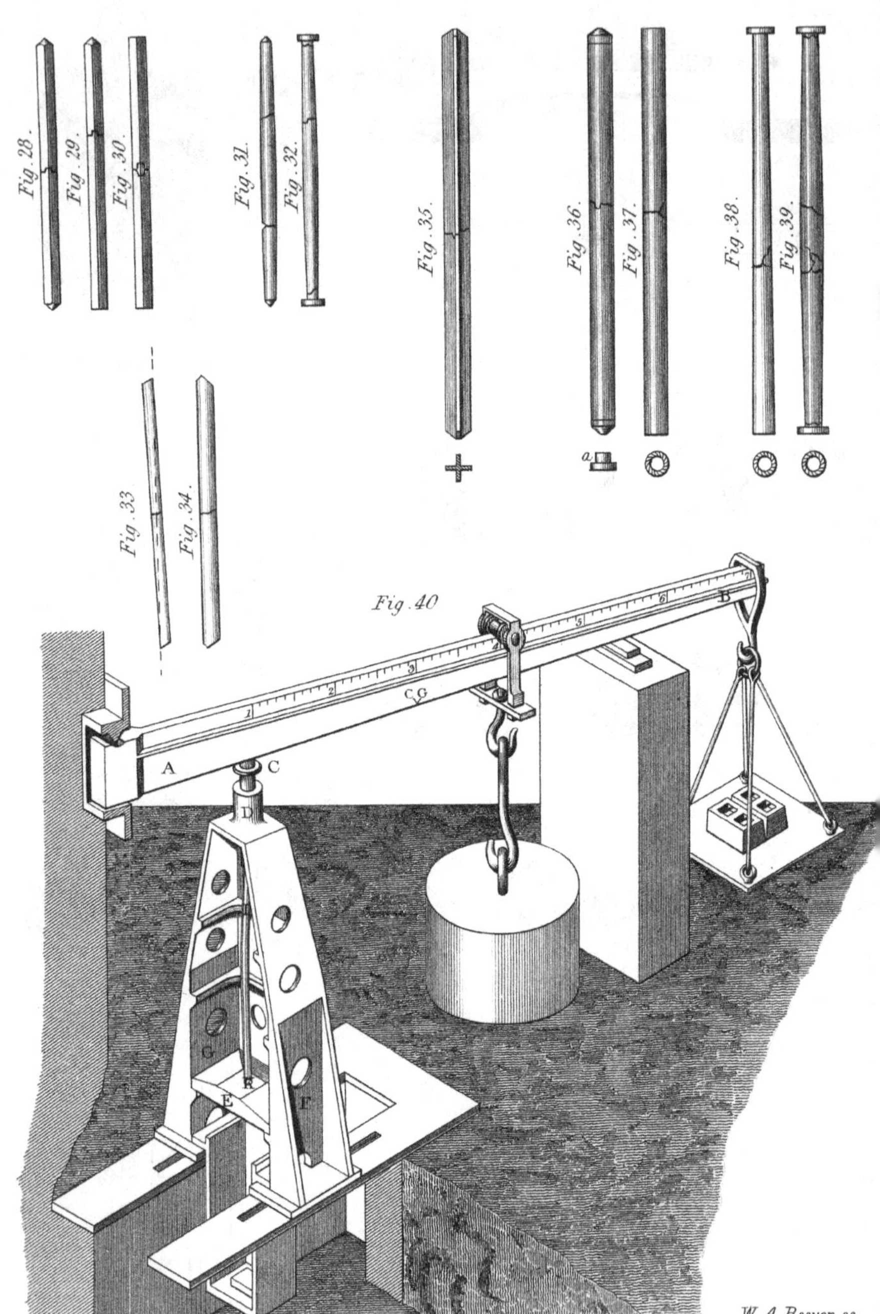

Published by J. Weale. at the Architectural Library, 59. High Holborn. 1846.

Plate. III.

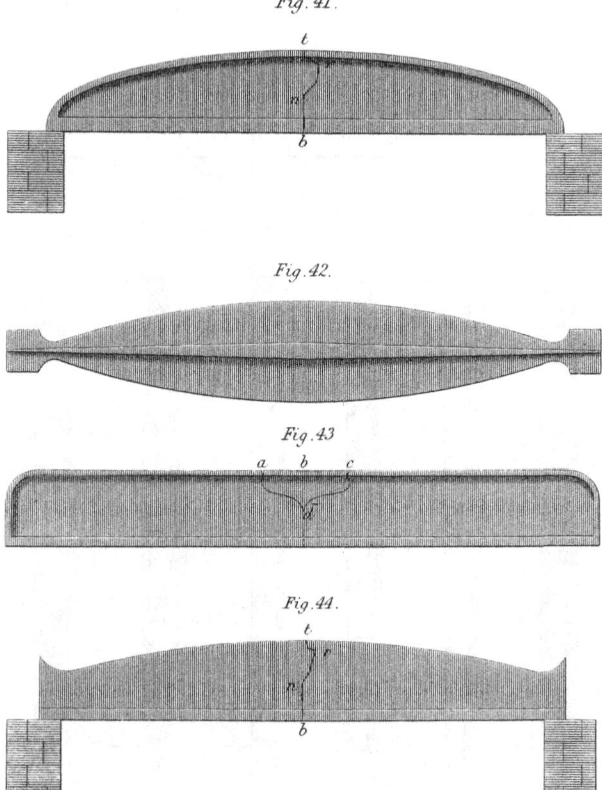

W. A. Beever. sc.

Published by J. Weale. at the Architectural Library. 59. High Holborn 1846.

Plate IV.

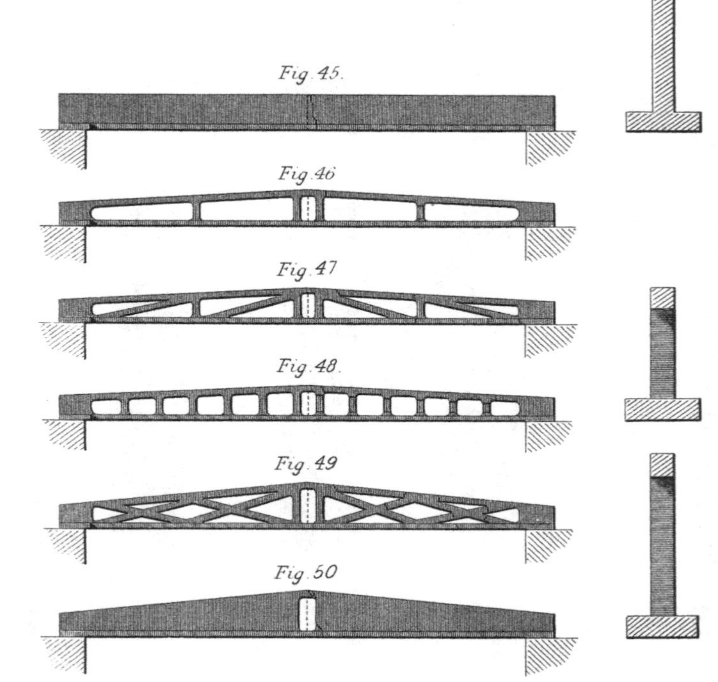

W. A. Beever sc.

Published by, J. Weale, at the Architectural Library, 59, High Holborn, 1846.

Printed in the United States
By Bookmasters